TRANSMISSION

TRANSMISSION

EDITED BY PETER D'AGOSTINO

TANAM PRESS NEW YORK 1985

DEDICATED TO
ANNUNZIATA D'AGOSTINO
1899-1984

FOR INFORMATION ADDRESS
TANAM PRESS
40 WHITE STREET
NEW YORK, NY 10013

PRINTED IN THE UNITED STATES OF AMERICA
ON ACID FREE PAPER
ISBN 0-934378-25-8(cloth)
ISBN 0-934378-26-6(paper)
FIRST EDITION

Preface

TRANSMISSION IS A SOURCE BOOK of new television theory and practice. It offers a broad framework for the study of television, one which encompasses television aesthetics, social commentary and applications of new technologies. The essays range in focus from expositions of communications theory and the social significance of mass media to the exploration of television as an art form.

An early reference point in my thinking about television is Walter Benjamin's "The Work of Art in the Age of Mechanical Reproduction." His critical analysis of film and photography, went beyond formal and representational meanings to investigate social, economic, and political implications of mechanically reproducible "works of art." This book seeks to extend some of these issues as they apply to the present "age of electronic transmission".

For ease of reference, *Transmission* is organized into three independent sections: theory, practice and distribution. However this is not intended to reflect limitations within the essays. On the contrary, many of the authors in this volume transcend the traditional categories employed in the study of television.

Rather than offer criticism on the usual TV genres of sit-coms, soaps and sports, I have selected essays that focus on programs that do not usually appear on commercial prime-time TV. Information on the work of Paik, Beckett, Kovacs, TV/TV and others may be more valuable in helping us to determine what television is, what it is not and what it may become.

Many significant programs which warrant critical attention only occasionally appear on television, others are never seen at all. One program which has never been aired on television offers an important definition of the television medium, both in aesthetic and socio-political terms. I am referring to "The War Game" produced by Peter Watkins for the BBC in 1965 and banned from the public airwaves even today. I want to bring attention to the serious limitations placed on significant programming, not only from direct censorship, but also from our habits of classification. Arts and cultural programs and others that can't easily be labeled, are generally excluded, by the public and scholars alike, from consideration as valid television forms. Ironically, many of these works are more intrinsically "television" than the so-called generic TV of sit-coms, police dramas, etc. that appear throughout the broadcast day.

My intention in assembling this particular set of ideas is to provoke new discourse on the interrelationship of form and content within the context of new television/video practice. The reader is encouraged to cross reference the essays, to compare the sometimes dialectical positions of the authors, to work toward a new synthesis of the ideas presented.

The word, 'transmission' refers to the technological apparatus of television, to the socio-political structures of the medium and ultimately to the exchange of knowledge. The publication of *Transmission* is one more step in the ongoing effort to redefine the map of television/video studies. A brief look at the bibliography and videography will show that this book is but one approach among many and that work is well under way.

* * *

For making this book possible, I want to thank the authors who generously contributed their work; Deirdre Dowdakin who helped with the organization and development of the volume; and Reese Williams whose initial support and continued encouragement led to the realization of the book.

Contents

PART II: PRACTICE

PART III: DISTRIBUTION

INTRODUCTION

Peter D'Agostino

Unlike all previous communications technologies, radio and TV were systems devised for transmission *and* reception *as abstract processes, with little or no definition of preceding content. . .*

It is not only that the supply of broadcasting facilities preceded the demand; it is that the means of communication preceded its content.[1]

In the beginning was the "S"

Although the development of radio and television broadcasting cannot be represented by just a single invention or event, Marconi's first trans-atlantic test of the wireless in 1901 (the message was merely an "S"—three dots in morse code) can serve to mark the beginning of the age of electronic transmission.

Wireless radiotelegraphy and later "radiophone apparatus" were two-way devices designed to send as well as receive telegraphic and voice information without the constraint of cable lines; they introduced an entirely new communications era. The growth of this new technology was, however, sharply limited by companies such as AT&T which were already providing two-way communications services of their own (telephone and telegraph). Sensing a threat to their monopoly, these companies were successful in preventing the use of the wireless for any applications other than "off-shore" maritime and trans-atlantic communications.

Ironically, this "imposed" limitation in the use of the wireless triggered the growth of an entirely new industry—radio broadcasting. "To shout the message. . . to receivers in all directions."[2]

Government regulations of the airwaves soon followed, and with the establishment of the Radio Corporation of America(RCA) in 1919, we see the beginnings of our present form of commercially sponsored radio and television. AT&T, one of four large corporations with a controlling interest in RCA, initiated the concept of "toll broadcasting" for RCA. This and

other forms of "ether" advertising started us on the path to the current situation: a one-way broadcasting system which is first and foremost a commercial enterprise.[3]

Broadcasting: "Sell-o-vision"

The invention of the television can be traced to early mechanical devices such as Nipkow's scanning disk for transmitting images (1884), but the complete development of a functioning television system is credited to both Philo T. Farnsworth and Vladimir Zworykin in the 1920's.

The rise of the television broadcast industry paralleled the development of radio with the "sponsor" firmly in control of programming. Some people have suggested that TV's "Golden Age" was in fact primarily a marketing strategy to sell TV sets—"sell-o-vision" it was called! Later, programs such as *Bonanza*, with its beautiful scenic locations, would be created to help market the new color TV sets.[4]

But, more importantly , as advertisers were quick to realize, television rapidly became the most efficient way to deliver people to consumer products. The medium became so powereful that entrepeneurs were able to use TV to create 'needs' for new products that had never existed before. In the "American system of broadcasting", television has become far more than a commercial enterprise, it is a marketplace of far reaching social, political and economic consequences—"a technology of cultural domination."[5]

Television and art

How does one consider commercial television and art in the same context? Can television (which is primarily thought of as a source of entertainment and news) be considered as an art form? These questions present a clear dichotomy of socially accepted but divergent notions of art: the elite "high arts" (traditionally painting , sculpture, music) as contrasted to the "popular arts" (in which television represents society's lowest common denominator). The confusion with regard to issues of "high" and "low" art has a long established history:

In fact, the division of art into autonomous and commercial aspects is itself largely a function of commercialization. It was hardly an accident that the slogan L'art pour art *was coined polemically in the Paris of the first half of the 19th century when literature really became a large scale business for the first time. Many of the commercial products bearing the anti-commercial trademark of art for art's sake show traces of commercialism in their appeal to the sensational, or in the conspicuous display of material wealth and sensuous stimuli at the expense of the meaningfulness of the work.*[6]

Is "art for art's sake" or the tautological definition of art "as art" purely a modernist tendency? Consider the following two statements:

"The signs of art must. . . bear a suitable relation to the thing signified."

"The medium is the message."

The first, by Gotthold Lessing, is from the "Lacoön, An Essay on the limits of painting and poetry" (1776) in which he articulates the premise that each art is imbued with its own distinctive properties, ones that limit its expressive and communicative function.

The second statement is almost too familiar. Although McLuhan carried these formalist principles too far, even for some formalists, they extend many of the concepts developed earlier by Harold Innis in "The Bias of Communication." Innis states, ". . . the unique power of each form alters the action of the other forms it encounters." Beginning with oral tradition, then continuing to the printing press, radio and television, he emphasizes the effects of specific communications technologies on social organization and culture.

Formalistic and technologically determined views such as the ones above have contributed insights, but they also have had a limiting effect on the study of television aesthetics. For instance, in *Sight, Sound, Motion*, Herbert Zettl reviews categories such as light, color, time and motion to show the principles of TV aesthetics. But there is absolutely no regard for content! The implication is that content is of little or no consequence when learning or practicing the art of television production.

On the other hand, the prevalent approach toward television in the social sciences has been to analyze content and then to study the effects on the audience. Although much has been learned about the impact of television's biases and misrepresentations, television continues to be defined outside the realm of art. And as a consequence, this research has had an insignificant role in changing dominant TV practices. (For an in depth study of content analysis, see *The Analysis of Communications Content* by George Gerbner et al.)

In our present post-industrial age, the explicit separations between form and content and between high and low art only serve to maintain the status quo of a consumer society's fragmented value system and world view. Their effect is one of disassociation at a time in history when many people are struggling in the opposite direction, toward integration and wholeness.

New perspectives are needed. Although the term "post-modern" is useful, in a larger view it is not sufficient to reconcile the "separations" mentioned above. We have some existing strategies—processes of deconstruction and appropriation of words and images in the public domain are providing an understanding of the framing mechanisms of a mass-mediated culture—and others will emerge if we orient ourselves toward the making of a television art designed specifically for the public airwaves.

Ideas "in the air"

In the past, this phrase was used to metaphorically represent the zeitgeist or spirit of the age. In 1985 we can now say, literally, that ideas are in the air as physical properties. Buckminster Fuller once described the phenomenon of broadcasting as having something to do with *tunings*: ". . . in the language of electromagnetics. . . I simply have what I have tuned in now and that's what we are conscious of, I can tune it out and bring it back again."

It's like the old cliche that somewhere, perhaps on Jupiter, an alien being is just now tuning in to Milton Berle. In a closed but constantly expanding universe, all programming "in the air" can return again and again for endless replays.

The overlap of imagination and reality is widely evident in the many visionary uses that have been proposed for communications technologies. Arthur C. Clarke, best known as the science fiction author of *2001*, was also the first to conceive and promote the concept of communications satellites (as early as 1945). But visionary propositions and technological advances must be tempered by the political realities of control and ownership of the public airwaves. The ideas and images now being transmitted "in the air" serve the political and economical interests of the communications industry rather than the interests of the public. Even the credibility of the Public Broadcasting System is undermined by its reliance on support from large corporate interests, including oil conglomerates and AT&T.

The burgeoning new technologies of cable TV, home VCR's, video discs and satellite distribution have temporarily diffused public concern over the centralized power of media networks by creating the impression that alternatives not only exist, but abound. However, the new technologies have failed so far to provide any real alternative and are being used only to propagate the values, goals and products of the mass media industry.

The fundamental question in "the age of electronic transmission" is whether television can begin to serve a broader constituency. How can telecommunication technologies be put to the service of people?

We must begin to imagine a future that is different from the one that is being subtly inficted upon us. The essays in *Transmission* contribute to this process in one of two ways. Either they challenge accepted beliefs about television, or they identify positive models for the future of television. My hope is that the ideas within in this book, when considered together, will create an 'interference pattern' within the seemingly continuous flow of television's prepackaged ideology.

Introduction to the essays

The theory section begins with Jon Bagaley and Steven Duck's essay *Hall of Mirrors*. Recounting a description of the BBC's first public television transmission in 1936, the authors explain how the medium seemed to provide a transparent 'window on the world' that became increasingly more opaque in the course of its development as an art form. They continue by investigating the psychological effects of television beyond traditional models of communication theory. The essay concludes with a look toward the implementation of an appropriately flexible and dynamic approach which is needed to understand the many levels of meaning conveyed in television.

Functions of Television is John Fiske and John Hartley's reading of TV from a semiotic and cultural perspective. They probe the relationship between the television message, the everyday reality of the audience, and the functions performed by TV for that audience. Outlining some of the dominant theories of how TV affects the audience such as 'dependency' and 'bullet theory,' they argue that TV functions in society as a form of communication: as a semiotic system of signs derived ultimately from pre-television (real life) linguistic codes.

In *Television Culture* Hal Himmelstein compares the popular art of television with notions of high art. He surveys the nature of mass culture and its sociological significance through the ideas of three leading figures in this debate: Dwight Macdonald, Edward Shils, and C. Wright Mills. Himmelstein then looks at recent analytic strategies focusing on TV as dream, myth, and melodrama concluding with comments on the significance of alternative artists TV.

John Hanhardt's *Watching Television* equates the role of the TV viewer with that of the consumer to define what it means to watch TV. He goes on to speculate what television could become in the future using radical models provided by artists' video in single-channel and installation formats.

The Whole World is Watching is Todd Gitlin's critique of the role of mass media as core systems for the distribution of ideology. Focusing on the complex relationship of television news and the new left of the 1960's Gitlin looks at the way the media frames actual events and deforms their social meaning. He analyzes the effects of these framing mechanisms on the world at large with regard to hegemonic ideology that attempts to naturalize the artificiality of media conventions.

In *...Visual Anthropology* to *The Anthropology of Visual Communication* Sol Worth introduces cultural issues in the study of communications through a description of the TV documentary series *An American Family*. He quotes from Pat Loud, the mother of the family depicted in the program who describes her feelings of how they were "ground through the big media machine." Worth goes on to show how film and TV do not depict truth but have the ability to present structured versions of knowing and seeing, illustrating his argument with examples from the work of Bateson and Mead in Bali and his collaborative project with John Adair, *Navajos Film Themselves*.

Cracking the Codes of Television is Howard Gardner and Leona Jaglom's study of how children organize the mediated world of television in relation to every day reality. They compare the young child to an anthropologist learning a new visual language and decoding the rules that determine the operation of TV. They also suggest that the aid of skilled informants in the guise of parents and teachers are essential in sorting out the confusion in the effects of TV conventions that blur a child's sense of reality and fantasy.

John Carey and Pat Quarles' *Interactive Television* surveys the brief history of the actual and simulated forms of two-way TV. They provide a detailed account of the Berks Community Interactive Cable TV system in operation since 1975 and designed primarily for use by senior citizens in Reading, Pennsylvania.

In *What is Videotex?* Vincent Mosco explores the international development of current data transmission systems connecting television-computer-phone lines, generically called Teletext and Videotex. He cautions that national policies on the production and distribution of information yields to market forces in shaping these systems in society. Mosco points out that while Videotex is promoted as a utopian technology it may in fact reflect the dystopian future of Orwell's telescreen.

Part II: Practice begins with *Ernie Kovacs: Video Artist*, in which Robert Rosen outlines Kovacs' accomplishments as TV's pioneer experimentalist. Unlike his contemporaries during the 'golden age' of TV comedy, Kovacs explored television's distinct aesthetic qualities rather than using the medium as filmed vaudeville or illustrated radio. He was broadcast TV's original innovator, exerting a lasting influence on the art of television.

Next, David Ross surveys *Nam June Paik's Videotapes*. The foremost video artist of the avant-garde, Paik is known as 'the George Washington of video'. Although his Fluxus 'neo-dadaist' style provides a distinct contrast to the dominant television industry practices, many of Paik's works were produced for and broadcast on public TV. It is Paik's major television productions such as *The Selling of New York* and *Global Groove* that are the focus of Ross' essay.

Samuel Beckett's Ghost Trio is Peter Gidal's detailed account of an extraordinary tele-play written in 1976. Conceived and structured by Beckett expressly for television broadcast, *Ghost Trio* contains three frames, one inside the other like a chinese box: the outer frame is the television screen itself, the next is that which is perceived in the visual narrative of the protagonist, and the interior plane is that which exists within his mind.

In *Nuclear Consciousness on Television*, James Welsh compares the recent commercial success of *the Day After* to Peter Watkin's 1965 production of *The War Game* which was banned from broadcast television by the BBC. Although almost two decades separate these two 'made for television productions,' a description of the controversy that surrounds

The War Game provides an illuminating look into the process by which broadcasters select and limit what appears on television.

Erik Barnouw's *The Case of the A-Bomb Footage*, chronicles the history of film footage of the devastation of Hiroshima and Nagasaki. Filmed by independent Japanese filmmakers, the footage was confiscated by occupation authorities and remained 'classified material' in Washington until 1968. After obtaining permission from the National Archives to review the films, Barnouw edited the material into the documentary *Hiroshima-Nagasaki, August 1945*. The essay concludes with a description of the television premiere of the program on the 25th anniversary of the Hiroshima bombing.

Guerrilla Television is Deirdre Boyle's look at some of the 'video documentarians' who created a new style of alternative television. Her primary focus is on the TV/TV group's innovative programs *Four More Years* and *Lord of the Universe*. She also speculates on reasons for the eventual demise of this alternative TV movement.

In *Meet the Press: On Paper Tiger Television*, Martha Gever discusses this current weekly cable-TV program. *Paper Tiger TV* examines dominant communications industry practices by providing alternative readings of the press. Produced by a collective, *Paper Tiger* provides a format for communications theorists, writers, and artists to confront the ideology of the mass media in the context of the industry itself.

Poet at Large: A conversation with Robert Bly is a transcript from the *Bill Moyers Journal* TV series. The program is especially noteworthy in revealing the inner complexity and paradox of poet Robert Bly's work, including his political commentary. Through a stimulating and highly informed dialogue, Bill Moyers provides a conducive setting to probe Bly's poetic imagination. The program can serve as a model for presenting poetry's oral tradition on television.

Part III: Distribution begins with *Tube with a View: Channel Four*, a brief overview of Britain's new TV channel by Kathleen Hulser. Beginning its first season in 1982-3, Channel 4's mandate was to provide a wide range of innovative programming, including independently produced women's and third world programs and experimental cinema.

The next three essays survey the three most important centers of experimental television in the United States. Marita Sturken provides a look at *WNET's Television Laboratory*, New York 1972-84; Joanne Kelly summarizes the activities of the *National Center for Experiments in Television, KQED*, San Francisco 1967-1974; and Susan Dowling chronicles the activities of *The WGBH New Television Workshop* initiated in 1974.

In *Video: A Brief History and Selected Chronology*, Barbara London summarizes video/television art of the past twenty years; the chronology includes exhibitions/events, television productions, organizations and publications.

Notes

For an in depth account of the history of broadcasting see *Tube of Plenty* by Erik Barnouw and *Stay Tuned* by Christopher Sterling and John Kittross.

[1]Williams, Raymond. *Television: Technology and Cultural Form*. New York: Schocken Books, 1975.

[2]Fahie, J. J. *A History of Wireless Telegraph*. New York: Arno Press, 1971.

[3]Barnouw, Erik. *The Sponsor: Notes on a Modern Potentate*. New York: Oxford University Press 1978.

[4]Hey, Kenneth. "Marty: Aesthetics Versus Medium in Early TV Drama," from *American History/American Television: Interpreting the Video Past* edited by John O'Conner, New York, Frederick Ungar, 1983.

[5]Schiller, Herbert. *Communication and Cultural Domination*. White Plains: Pantheon(dist.), 1976.

[6]Adorno, T. W. "Television and the Patterns of Mass Culture" from *Mass Culture: The Popular Arts in America* edited by Bernard Rosenberg and David M. White, New York: The Free Press, 1957

TRANSMISSION

PART I: THEORY

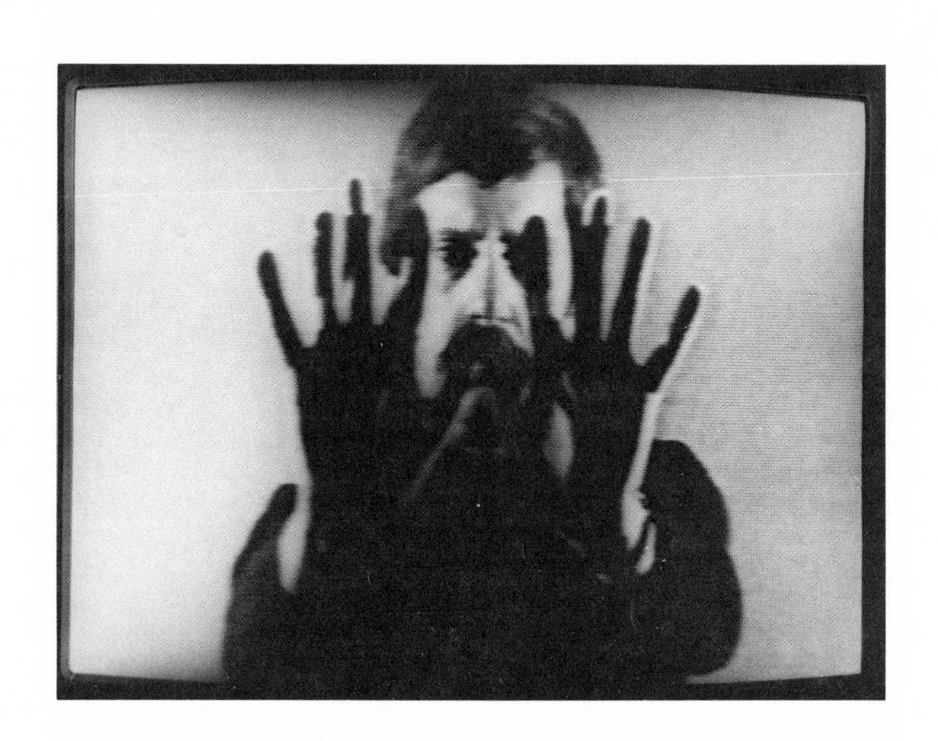

HALL OF MIRRORS

Jon P. Baggaley and Stephen W. Duck

A FAVORITE SIMILE for television is the sheet of glass. The medium's chief attraction lies in its properties for revealing a panorama beyond our immediate horizon; for reflecting a world of possibilities that would otherwise be denied us. We are with reason often alarmed by the possibilities television conveys, and the medium itself is a common object of blame for the social problems it brings to light. But we know little of the ways in which television actually affects us. We assume it does, for it certainly absorbs our time: but so, of course does sleep, and during sleep some of the most gross distortions of all are prone to visit us. So what is television's impact? Does it actually have the all-pervasive role in our lives that has frequently been supposed? Or, like dreams, does it meet a psychological resistance that tells us the reflections are conjured? In this essay we examine television's reflecting properties, we consider the intentions and techniques underlying the images it conveys, and also the ways in which its viewers attribute significance to the images they perceive. We also make a perhaps surprising but insistent point: that the subject matter conveyed by television is of less basic importance than is generally assumed; it is on the viewers' reactions to the imagery of television presentation that we must concentrate if the medium's fundamental psychological effects are to be established.

The transparent medium?

On 30 October 1936, the world's first public television service was launched from London's Alexandra Palace. The first evening of transmission featured a special BBC film, *Television comes to London*, to mark the historic occasion. The *Radio Times*—extended to include the new television schedule, and still known under its old name today—described the film as follows:

[It gives] an idea of the growth of the television installation at Alexandra

Palace and an insight into production routine. There will be many shots behind the scenes. One sequence, for instance, will show Adele Dixon as she appears to viewers in the Variety at 3:30 this afternoon, and will then reveal the technical staff and equipment in the studio that made this transmission possible. (Graham 1974, p. 14)

It was doubtless of no surprise to the public of 1936 that the technology of television projected an image as potentially opaque as a scene in a theater, behind which numerous unseen processes were operating. In revealing its behind the scenes secrets on the very first day of transmission, television also displayed a glorious ingenuousness. For what, at that stage, did its mysteries really consist of? In 1936, television was seen exclusively as a "window on the world." During the inaugural week, the TV program schedule included concerts, a flower show, demonstrations of boxing, tap-dancing, and electronic music, though no actual use of the medium as an art form in its own right. Drama and ballet were featured, though each was relayed from one of the London theaters. The relative artlessness of television's own projection styles at that time represented a translucency that was soon to be lost as the art became more sophisticated. In the early days of television, enthusiasm for the medium's simple novelty generated production practices that today seem touchingly unadventurous. The evocative "interlude" presentations, for example (*Waves on the Beach*, *Kittens at Play*, *Storm-tossed Clouds*, *The Potter's Wheel*): twenty years ago, these forms were acceptable (and, dare we say it, would be so again now—as the recent imitation of this form by certain British advertisers indicates). The medium *was* transparent. Its very transparency as a window on nature was exalted. But after the aesthetic value of a medium's simple novelty has declined, it must develop other means for reinstating itself in its audience's eyes. Gradually, the transparent medium of television became more opaque as the stylistic inventiveness of its users developed; and the couse of this development is the same in all artistic media.

In the early days of film, the inventive play was even more transparent than it was in television. As a totally recorded medium, film is far easier to manipulate than its live counterpart, and the early work of the Lumière brothers in France at the turn of the century testifies to the boundless and magical possibilities that the medium was seen to offer. It is only in the 1970s that the electronics of television have become sufficiently sophisticated to compete on the "special effects" market also. In celebration of the chroma-key technique alone a fabulous style of documentary presentation has become possible in which musical works and the creative language of all the arts are expressed in multiple images, freely fusing and dispersing. The potential of television as an art form in its own right is now in the process of realization; and since art is essentially the antithesis of transparency (witness the objections of some critics to pointillism as art), "opaqueness" is now a necessary ingredient of TV stylistics as in any art form. (But consider our suggestion that a return to the simplicity of *Kittens at Play* might not be as unwelcome as one might think!)

Like any art, therefore, television is now skilled in the art of illusion. The skills of the producer, the actor, the film editor, the cameramen and engineers combine to create a product that will satisfy the criteria viewers have learned to apply to it. Occasionally, the behind the screen secrets of television production are partially revealed anew, as in such introspective programs as the BBC's *Inside the News* and *Looking at Documentary* (both 1975) and *Film as Evidence* (1976): indeed, the recent introspectiveness of the British TV networks in this respect is an interesting feature of the current broadcasting climate, in which the Annan Committee is sitting to discuss the future of network television over the next few years. But although the techniques and mechanics of the medium are sometimes acknowledged, they are rarely allowed to become apparent in the creative act. Unless a particular style of media presentation is employed for artistic effect in its own right, the microphone and camera shadows are kept carefully out of the picture. For the medium to intrude upon the action is felt to destroy the illusion it creates. However, the presentation criteria that the medium applies in treating its material do not apply in the presentation of television's overtly artistic content alone. As television learned to emulate the traditional emphasis of the cinema upon constant variation of the image and the smooth continuity of its elements, those rules were applied in the presentation of drama and current affairs alike. The early criterion that newsreaders should be heard but not seen had, by the 1960s, given way to the norm by which they could not only be seen and heard but also identified by name. During the 1970s the television news on both sides of the Atlantic has been presented by two newsreaders per program, alternating in the interests of visual appeal alone. Their appearance on the screen is buttressed by charts, photographs, filmed insert material (often with little relevance to the action reported), and gratuitous shots of the newsroom in which they work. The news is packaged and promoted like the soap operas and adverts that follow it; and the decision by the BBC news editor in 1976 to cut down on purely irrelevant packaging, even returning to a solo reader, has created as much interest on the part of the other news media as the news material they collectively convey: "the news on BBC is going to lose its trendy image . . . [dispensing with] many of the gimmicks that marked the presentation factors of news during the past year . . ." (*Daily Mail*, 19 February 1976).

The role of presentation in the television context may be seen as essentially similar to that of the elaborate non-verbal behavior practiced in communication situations generally (nods, smiles, eye movements, etc.). Like any mediating glass, television can distort and transform the material it presents by the way in which it presents it; and whilst it would be misleading to suggest that the non-transparency of a medium necessarily leads to dishonest reflections of its subject material, the means of presentation is certainly an important influence on the message's reception. Any message, in order to be communicated effectively by television or any other means, must be couched in terms that attract interest; and the techniques employed for this purpose in human communication generally are in no way exclusively verbal; without the underpinning of inflections, facial expressions, posture and an elaborate routine of nods and non-verbal signals perfected through generations of practice, the immaculate verbal logic of utterance is to no avail (cf. Argyle 1969). An effective range of non-verbal skills is essential to the communication process, for by it the message stands or falls. In its absence, individuals are unable to form the basic human relationships from which communication stems; and in their attention to it, moreover, individuals are subject to the most unpredictable of prejudices.

The prejudicial ways in which a message may be interpreted are a central issue within this book. In normal conversation, our attitude to every word that is spoken is profoundly influenced by the relationships of trust and expectation by which the interaction is colored; and if the figures that address us via television are to command our attention in the normal way, they must naturally be subject to the usual laws and conditions. On television, a succession of people claiming authority on one subject or another appears before us each day. If we have no prior knowledge of them, it is necessary for us to form an impression on whatever immediate basis we can: an impression as to whether they are sincere, humorous, and so on. For no one likes to suspend judgment until the case has been argued: and, with the evidence of both eyes and ears, few can. When we speak, therefore, of a person's television "image" we refer to a combination of ideas and predictions about that person which may vary according to the values we each apply. When we speak of a public image—either of an individual or of a cultural group—we imply that the audience's general reaction on the subject is united: and this indicates either some prior knowledge or a deliberate and powerful projection by the image makers. The image of a person, like a caricature, is a selection and amplification of certain features at the expense of others; and on television the nature of the impression a performer makes is potentially determined not merely by the performance itself but also, as we have indicated, by those who mediate it, by studio interviewers and by the complex television technology that the studio staff manipulates.

The extent to which television gives an actually dishonest reflection of the material it conveys is thus, in part at least, related to the intentions underlying the image building process. But distortions also occur when any aspect of the presentation is misinterpreted or gives rise to unwarranted assumptions. Not only is the performer's personal image liable to be deformed: the meaning of his spoken word may be misshapen to fit it. Whether it is formed consciously by the performer and his mediators, or assembled by the viewers according to their own criteria, the performer's media image may even determine the amount of basic attention paid to him. In ordinary human interaction, we accept that little of the speech passing between us fully registers. We may assume that others understand our spoken intentions, but would admit that the attention we pay in return is less than perfect. Based on non-verbal skills alone, a satisfactory human exchange may occur with little concentration on what is actually spoken, and with the best of intentions two people can maintain the illusion of communication for hours. Though we accept this as a feature of face to face encounter, we forget it in our analyses of man and the media.

For media research has hitherto been based on the fatal assumption that a medium's influence may be analysed in terms of the effects of its subject matter alone. Media practitioners, on the other hand, realize that if visual presentation is neglected and the rich non-verbal code of cuts, fades and superimpositions is dispensed with, then audience attention will quite simply flag. The low esteem of much media research upon media practitioners is thus hardly surprising. Emmett (1975) has summarized the broadcasters' often resentful attitude to media research in the following terms: their irritation with what they feel to be gratuitous advice, their annoyance at the underestimation of their genuine concern for media development, and their scorn for the often elementary nature of media research. But this is not to suggest that broadcasters scorn research *per se*: indeed, the major broadcasting organizations mount an expensive audience research effort. It is quite likely that focused specifically upon the effects of production art itself, research techniques might reveal effects other than the producers have intuitively realized: and these, as indicated by Robinson and Barnes (1976), actually would be of interest to them.

But if one's powers as a communicator are affected by the image that is presented, and if media presentation in turn is ruled by superficial criteria for simple variety, how much of televised content ever actually makes any impression? Previous research based on the analysis of subject matter alone has assumed that the latter comes over to its viewers in more or less pristine fashion. But does television actually speak to its audience with any more credibility than any other human agency? And does the audience really heed all the information it imparts? Or, like the man who plays the game of communication with us, does the medium fulfill most of our needs at a lower level than we like to imagine? We may imagine ourselves to be assimilating a program's argument, only to realize, by the lack of any subsequent recollection of it, that we have happily spent the hour otherwise. Hearing our name spoken elsewhere in the room, we may recall the immediate context in which it occurred but have no recollection over the longer term. So context does go in, it seems, but swiftly decays unless we have reason to intellectualize it. And in our preoccupied state in front of the television set, it follows that the logical content of the transmission may often be of less significance than the imagery that mediates it, the shadows and transitions of presentation alone.

In the literature of television research evidence regarding presentation effects is sparse (Coldevin 1976). Anderson (1972) reviews an abundance of "non-significant" and inconsistent effects elicited when researchers have tried to compare the actual instructional effects of TV presentation (e.g., color, motion, focal length, and lighting effects); and, admittedly on intuitive bases, he is led to suggest that research into the effects of presentation on attitude and interest levels may reveal a dimension on which the instructional gain from television is dependent. The educational programs produced in the US by the Children's Television Workshop (Palmer 1969) are accordingly based on the premise that the power to instruct depends on the ability to attract attention via presentation strategies specifically. Their classic series *Sesame Street* employs a wide range of studio and location techniques, music, humor, puppetry and graphic work to teach the principles of literacy and numeracy to children in the lower junior school range. (It is also hugely entertaining to adults.) *Sesame Street's* successor, *The Electric Company*, has been designed on the same principles, using, for instance, song and dance and sketch material with success, though it has achieved slower international sales. The third series from the CTW, using similar techniques, has been *Feeling Good*, which provides instruction on personal health for adults. This, however, has been the least popluar of the CTW products, though the reasons that the series does not "feel so good" as the others are not readily apparent (Palmer 1976). What the effects of these series do show, however, is the dynamic development not only of TV techniques but also of audience responses to them: a well-principled technique is only effective for so long before the audience itself becomes sophisticated in its response to the genre and demands something a little different.

The literature of the cinema, with a far longer history, has been deeply aware of this and in film analysis the questions of direction and artistic technique are uppermost (Wollen 1969). The order and context in which particular shots and slices of the action are presented has long been known to be of prime importance in the manner in which it is interpreted (Eisenstein 1947); and Isenhour (1975) reviews a collection of experiments conducted by Eisenstein and other pioneers of film direction and analysis such as Goldberg (1951) and Pudovkin (1958). The Kuleshov-Pudovkin effect reported by Pudovkin indicates that a neutral shot will be interpreted in dramatically opposite terms according to the shots with which it is juxtaposed: thus an expressionless face seems either happy, pensive or sad according to the shot by which it is preceded (e.g., a little girl playing, a bowl of soup, or a dead woman in a coffin). Whether any material conveyed by the visual media can ever appear totally unambiguous is therefore extremely debatable. Some writers on film, notably Bazin and Barthes (also cited by Wollen 1969) have maintained that a clear expression of its intended meaning is possible, and that the intentions of a medium to reflect "real events" and scenarios can thus be transparent; but this is only the case if the audience is well-versed in how to interpret the medium by a process akin to visual literacy—and, moreover, if it has the opportunity to check its interpretations. It is a conclusion to which have been led by the data we have observed, both experimentally and from previous writings. We conclude that the "transparency of the medium," as often upheld, is a myth put about by those who do not appreciate its psychological subtleties. A variant on the same myth supposes that freedom of access to the medium will effectively demystify its subtleties, rendering it once again as transparent as the day it was born. We argue, however, that the basic properties of television relate to no fixed and finite set of factors but to the dynamic psychological processes whereby its content is construed. We suggest that in educational use, for example, an understanding of these processes may be harnessed in the interests of greater control over the medium's informative properties in future.

Traditional theories of communication, however, provide little clues as to either the psychological properties of media or the possibilities for their more controlled usage. In the following section we therefore examine traditional models of communication and relate them to the views held by philosophers of man the perceiver of the world around him.

"Signal and noise"

The purpose of communication theorists (cf. Lin 1973) has been to define the processes involved when a signal, or message, passes from its source to an independent system with facilities for signal detection. At the simplest level of understanding, the communication act may be described as in Fig. 1, where "to communicate" is seen exclusively as the act of the message sender, while the recipients, totally passive, merely receive. In 1949, however, the model was described rather more precisely by Shannon and Weaver; by direct analogy to the electrical engineering model of communication, they added to the basic view components of message transmission and decoding (see Fig. 2). In this model the recipient is less docile: on leaving its source, a message is conveyed in coded form as a signal to a system capable of deciphering it. Within the channel used, however, communication may be impeded by interference with the signal, by "noise": thus the communication process is acknowledged to be fallible.

But the model is still deficient. The origins of "noise," for example, remain inexplicit; moreover, as indication that it has been communicated successfully, the independent system's fidelity to the signal in relaying it must be examined. Only then is the extent to which communication has taken place to the required degree strictly apparent—for it may emerge from this analysis that the transmission of the message has been influenced by "noise" factors not previously taken into account.

A more explicit version of the Shannon and Weaver model is thus presented by Fig. 3. The communication linking two participant systems is seen as a circuit, in which a message is dispatched from its source, coded, transmitted, received, decoded, and ultimately relayed by the same process. The destination of the intial message becomes in turn the source for its relay; and the measure of communication is the extent to which, despite the impinging noise within the system, the message can be returned to its original source in its original form. The main advantage of this expression of the process over the traditional engineering model is that noise factors are recognized as arising at any stage in the circuit and not from external agency. Several theorists have emphasized the fallibility of human communication, placing different stress on its origins at various stages of the process. Burke (1945) indicates the complexity of the human transmission process in communication, and the effects of the message sender's underlying motivation. Lasswell (1948) defines the variables underlying communication more fully as "who says what in what channel to whom with what effect?"; while other writers (e.g., Riley and Riley 1959; Secord and Backman 1974) stress the dependency of the communication process on the social system in which it occurs. Our control of communication processes depends on the extent to which all of these extraneous effects in their various forms can be identified and minimized. In our analysis of effects arising when a message is mediated via television, therefore, we seek to distinguish signal from noise not only

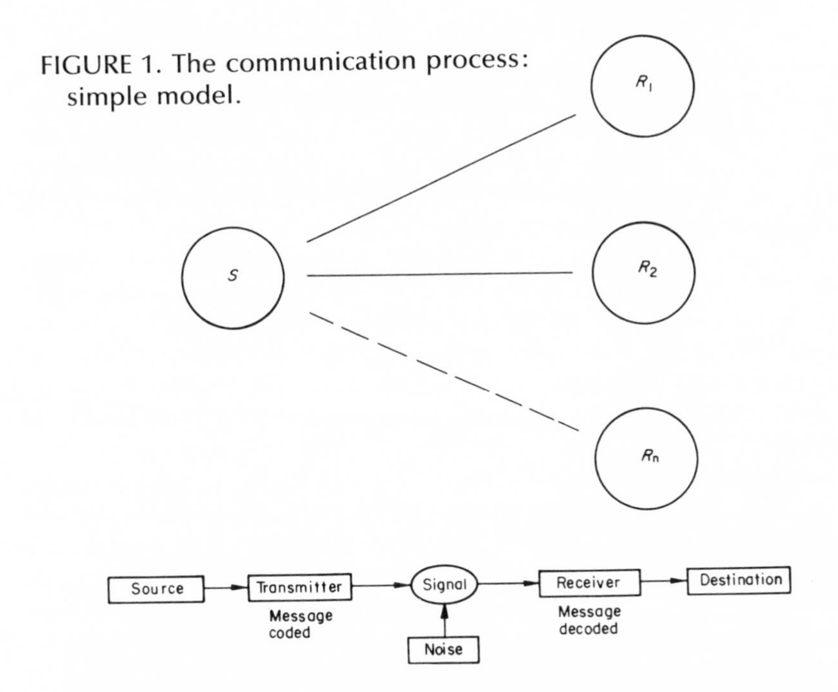

FIGURE 1. The communication process: simple model.

FIGURE 2. Shannon and Weaver's(1949) communication model.

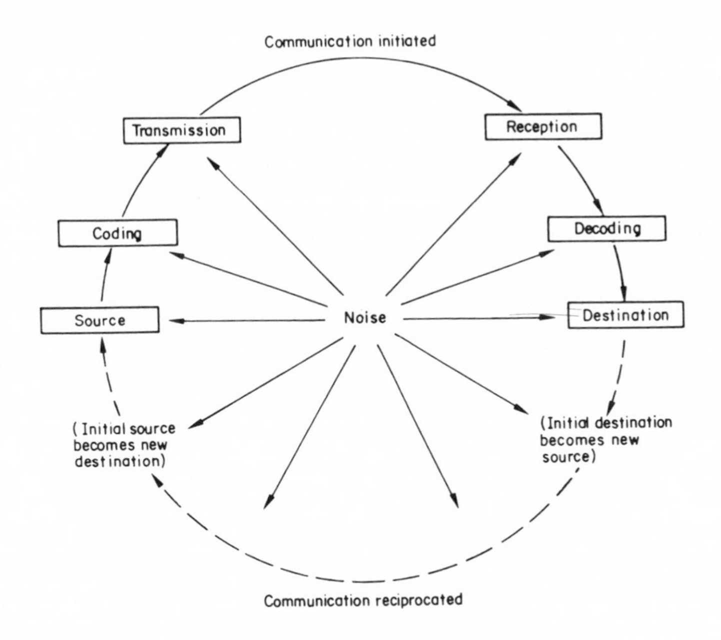

FIGURE 3. The communication circuit.

in the nature of the source and the coded form in which the message is expressed, but also in the viewer's receptive capacity, and the logic used when the message is encoded. The effects of "noise" may never be cancelled entirely, though as one's capacity for separating the noise from the signal increases, so the more severe problems of communication in a given medium can be avoided.

In electronic systems of communication, the cause and effects of noise can now be controlled with some sophistication. But since noise factors in human communication processes are not merely technical but also psychological in origin, our control over these is more limited. We need to identify the dynamic effects of attitude and experience on the ways in which the human system attributes meaning to the message it receives, meaning that may never be explicit solely in the form a message takes on re-transmission. We must recognize, however, as in the previous section, that the mediated form a message takes may in fact provide the only definition of its content there is: that little intended meaning may actually underlie it. And in view of these problems, a theorist might be forgiven for doubting whether clearcut definitions of the communication process is ever possible!

This thought has often occurred to philosophers. A fundamental question of interest within philosophical theory has been that of human perceptual ability, without which little successful adaption to the environment is possible, and certainly none of the two-way communicative type. The question has been posed in two traditional forms:

1. What are the actual practical *mechanisms* that man employs in perceiving the world around him? and
2. What is perception? Can man ever, in fact, really know what he perceives at all?

Before coming to be identified with distinct traditions within psychology and philosophy specifically, these two forms of the question were classified as *physical* and *metaphysical*. A problem illustrating the two approaches is that of visual depth perception (Boring 1942). To the philosophers the puzzle was in understanding how an object may be seen as three-dimensional even though the image of the object on the retina is two-dimensional. Until the seventeenth century, the basic theory of the physicists in this respect was that an object's distance is perceived strictly according to the invisible angles at which light waves from an object converge upon the two eyes separately. The British empiricist school of philosophers in the seventeenth and eighteenth centuries espoused the notion of the physicists that all knowledge derives purely from the senses in this way. Locke indicated that the senses convey to us information about an object, some of which accurately represents the reality of the object and the rest of which does not. Berkeley went a stage further than this, and was more realistic than Locke, by pointing out that we never can know for certain the extent to which this is so—the extent to which the real world and our mental images of the real world coincide—for we only have our mental images to go on. So saying, he in effect restated the argument put forward by Plato in the fourth century B.C. that nothing in the world around us is *really* real at all. To Plato the world of phenomena is a poor reflection from another ideal world; in his theory of forms he offers the "cave analogy" which suggests that man exists within the confines of a cave into which the *real* world casts shadows. All that we can perceive is a reflection, an interpretation of reality as in a mirror. (The mirror is thus our opaque medium, and the effects of the world upon us can only be understood in terms of the ways in which we interpret it.) The idea is reiterated by the modern philosopher Popper (1963), who cautions us that, while as scientists we may search for the truth, it will always ultimately elude us: "for all is but a woven web of guesses."

The cave analogy is at once a great statement of the fallibility of human processes and of their psychological complexity. It counsels a modest realization of our inability ever to formulate the final answers to our questions; but it indicates that the most satisfactory answers may nonetheless be sought from several different directions at once. The processes of human perception, and thence communication, may be studied by an examination of the mind's interpretive processes in a variety of ways; and Berkeley's contribution to the problem of visual depth perception was to consider the ways in which the mind *organizes* the material the senses convey to it accordingly. For, asked Berkeley, if the perception of depth depends upon the simultaneous agency of both eyes at once, what about the one-eyed man? He can perceive the three-dimensional properties of the world despite his disability, so the strictly physical theory of light waves and angles must be inadequate. In part at least, the mind must interpret what it perceives according to the ways it has *learned* to do so.

Within twentieth century psychology, the different ways in which individuals interpret their world have been emphasized in numerous behavioral contexts. Bartlett (1932) spoke of the set of psychological *schemata* governing individual behavior; Gregory (1970) has referred to the process by which information is judged in relation to a set of *internal models*. Kelly (1955), on the other hand, has propounded a theory of *personal constructs*, indicating that man views (construes) the world according to his own characteristic set of psychological criteria. Taken at this level, of course, such theoretical notions merely duplicate those for which the Greeks also had a word, when they referred to mental *traces*.

But, in fact, Kelly's theory goes further than merely describing: it aims to understand the process by which man relates his percepts to his behavior in everyday life. Man in general, argues Kelly, uses the rational approach of a scientist in viewing the world, gathering evidence about it, forming hypotheses and testing them out. Constantly relating observations to his repertoire of personal constructs, he analyzes the complex interaction of variables that he perceives and thereby comes to terms with the environment in which he must function. Any process by which one item of information is related to another, Kelly indicates, involves their comparison for similarities and differences; and in Kelly's personal construct theory the analysis of incoming data is seen explicitly in these terms. The theory provides with an insight into man's psychological processes which is independent of the type of determinist assumption that has mitigated against psychological theory in the past.

The philosophers' view of human perception as a glimpse, through a clouded medium, at absolutes that may never be defined can set our mind at rest as to whether we shall ever understand the total impact of external forces on human cognition *in vacuo*. Yet previous attempts at the analysis of the media's impact via definitions of its subject matter seem to have assumed that this is possible. In the total framework of research into television's effects, an analysis of subject matter is an indispensible element; but equally indispensible is the role of he who receives it, the viewer. In the impersonal nature of the broadcast medium at present, unfortunately, the viewer's role in the process is rather more assumed than known. (Only in a two-way communication system is it actually made plain.) But, as we shall attempt to show, it makes little sense even so to study the medium in terms of its message input alone. Even if a strict analysis of the television message were possible we should see that its journey from source to destination takes it through a hall of mirrors with numerous dynamic properties for dissolving and reconstituting the meaning it conveys.

Media dynamics

The essence of dynamism is change; and a question as many-sided as the impact of television demands an approach that caters both for the changes that take place within the medium itself—on the part of those who design its content—and for those on the viewers' part that determine its final efects. Thus we examine the shifting criteria underlying the production of the television image, the psychodynamic factors determining the viewing experience, and the extent to which, despite all this flux, the effects of television can be identified and controlled. Central to each aspect of the question is the "television message," concerning which the biggest assumptions of all have been made hitherto. In the view of the total flux apparent at every level of the human communication process, the communicated message is the very element that cannot be defined *a priori*. We may assume the nature of the message for simple descriptive reasons, but may base no conclusions about its cause or effects upon these assumptions. Moreover, if we wish to predict the effects of a particular message on another occasion, on another audience, and in another context, we must be prepared to find that our existing description no longer obtains.

Our initial frames of reference, then, must be capable of revision and even rejection according to the situation and moment under analysis. The descriptive approach we adopt must avoid any assumption that the properties are ever discrete and finite—we must also avoid all labels that in themselves attempt to pin down the phenomena too fixedly. We must look not merely at the phenomena themselves but for the forces behind them. By observing the relationships between the various points we use, however, we may ultimately come to understand the dynamics of television and be able to predict them.

Summary

(1) Television has been seen as a medium capable of reflecting the world at different levels of accuracy. The role of presentation in the communication of a message, however, invariably affects its impact; and the over-emphasis of previous research on the effects of television's overt subject matter needs redressing.
(2) Models of the communication process, attempting to define its variability in terms of static "noise" factors, inadequately reflect its psychological complexity. Ancient and modern philosophers remind us that psychological capacities are essentially fallible and determined by the dynamic processes of organization and experience.
(3) In order to understand the effects of television we must examine its dynamic properties via an appropriately flexible approach.

References

Anderson, C. M. (1972), "In search of a visual rhetoric for instructional television," *AV Communication Review*, vol. 20, no. 1.

Argyle, M. (1969), *Social Interaction*, London, Methuen.

Bartlett, F. (1932), *Remembering*, Cambridge, Cambridge University Press.

Berkeley, G. (ed. D. Armstrong), *Berkeley's Philosophical Writings*, London, Collier Macmillan, 1965.

Boring, E. G. (1942), *Sensation and Perception in the History of Experimental Psychology*, New York, Appleton-Century-Crofts.

Burke, K. (1945), *A Grammar of Motives*, New Jersey, Prentice-Hall.

Coldevin, G. (1976), "Comparative effectiveness of ETV presentation variables," *Journal of Educational Television*, vol. 2, no. 3.

Eisenstein, S. M. (1947), *The Film Sense*, New York, Harcourt and Brace.

Emmett B. (1975), "The perspective of the broadcaster," paper read to the Conference of the British Psychological Society (Social Section), University College, London.

Goldberg, H. D. (1951), "The role of 'cutting' in the perception of the motion picture," *Journal of Applied Psychology*, vol. 35.

Graham, B. (1974), *Television*, London, Marshall Cavendish.

Gregory, R.L. (1970), "On how so little information controls so much behaviour," *Ergonomics*, vol. 13.

Isenhour, J. P. (1975), "The effects of context and order in film editing," *AV Communication Review*, vol. 23, no. 1.

Kelly, G. A. (1955), *The Psychology of Personal Constructs*, New York, Norton.

Lasswell, H. D. (1948), "The structure and function of communication," in Bryson, L. (ed.), *The Communication of Ideas*, New York, Harper.

Lin, N. (1973), *The Study of Human Communication*, New York, Bobbs-Merrill.

Locke, J. (ed. A. C. Fraser), *An Essay Concerning Human Understanding*, New York, Dover Publications, 1959.

Palmer, E. L. (1969), "Research at the Children's Television Workshop," *Educational Broadcasting Review*, vol. 3, no. 5.

Palmer, E. L. (1976), keynote address to the International Conference on Evaluation and Research in Educational Broadcasting, Open University.

Popper, Sir K. (1963), *Conjectures and Refutations: the Growth of Scientific Knowledge*, London, Routledge and Kegan Paul.

Pudovkin, V. I. (1958), *Film Technique and Film Acting*, London, Vision Press.

Riley, J. W., Jr. and Riley, M. W. (1959), "Mass communication and the social system," in Merton, R. K., Broom, L. and Cottrell, L. S. (eds.), *Sociology Today: Problems and Prospects*, New York, Basic Books.

Robinson, J. and Barnes, N. (1976), "Evaluation in adult education broadcasting?" *Journal of Educational Television*, vol. 2, no. 3.

Secord, P. and Backman, C. W. (1974), *Social Psychology*, New York, McGraw-Hill.

Shannon, C. E. and Weaver, W. (1949), *The Mathematical Theory of Communication*, Urbana, University of Illinois Press.

Wollen, P. (1969), *Signs and Meaning in the Cinema*, London, Secker and Warburg.

A Vast Waste Land

THE FUNCTIONS OF TELEVISION

John Fiske and John Hartley

Just as all living organisms live in certain specialized environments to which they adapt and which completely determine their lives, so do human beings live, to a significant extent, in an ocean of words. The difference is that the human environment is, to a large extent, man made. We secrete words into the environment around us just as we secrete carbon dioxide and, in doing so, we create an invisible semantic environment of words which is part of our existence in quite as important ways as the physical environment. The content of verbal output does not merely passively reflect the complex social, political, and economic reality of the human race: it interacts with it as well. As our semantic environment incorporates the verbal outputs secreted into it, it becomes both enriched and polluted, and these changes are largely responsible for the course of human history.[1]

Rapoport's thought-provoking metaphor suggests that we both create and are sustained by our language. We are, in other words, produced by the environment of signification that we have collectively produced. Part of that environment comprises the constant stream of "secretions" that emanate from the small screen. However, we cannot merely "ingest" those secretions, any more than we can merely ingest food. Just as our metabolic processes transform what we eat into material that can be assimilated, so our culturally learnt codes and conventions transform what we watch from mere external stimuli into actual *communication*, where the message is not only received but also decoded, understood and responded to.

Moreover, as Rapoport's metaphor indicates, the codes and conventions which comprise our particular culture's ways of seeing are incorporated into the modes of perception of each individual to such an extent that we are largely unconscious of their operation, just as we have little consciousness of, or control over, our metabolic processes when we digest food. Our perception is not so much an inherited mechanism as a learnt one—the daily manifestation of our whole personal history of socialization and interaction with the cultural environment. Hence the awareness we bring to the television screen is a precondition for making sense of what we see, but that awareness is itself produced in us by what we have experienced hitherto.

Of course it is often difficult in practice to bear this complex relationship in mind. After all, television is literally a highly visible medium, and it does seem to influence people's behavior, if only to the extent that more people watch for more hours today than they did a generation ago. It is a short step from this observation to one which proposes that television, unaided, *causes* people to sit and watch. It comes in some quarters to be regarded as a pest. It has the same kind of reputation as a fox who gets into a chicken coop and kills indiscriminately far more chickens than it could possibly eat or drag away. Clearly the fox is a menace and one of nature's deviants. However, in the heat of the moment it is easy to forget that the fox exhibits this wanton behavior not through natural malice. On the contrary, his behavior is conditioned by the environment which the farmer has created, whereby more chickens than would naturally congregate together are cooped up with no means of escape. The "historical circumstances" governing the lives of the chickens are the cause of their downfall, not the fox. But the onlooker finds it convenient to blame the innate brutality of the predator, and not the farmer's culturally determined method of keeping chickens.

Similarly, television is often blamed not so much for killing chickens—or even viewers—but certainly for producing results which are in fact conditioned by much broader and more diffuse historical circumstances. Even the number of hours people spend watching television is not ultimately caused by television itself. It results, much more obviously, from shorter working hours, increased family resources available for leisure, and from the likelihood that in many cases the hours television fills were previously filled by various activities, like knitting, chatting or even dozing, with which the new medium seems to be able to co-exist quite comfortably. It has brought new stimulus into the home, and created a demand for more rather than less entertainment of other kinds. More books, magazines and newspapers are read, more music heard, and more plays and films are seen now than ever before—even if "only" on television. It brings a good many of these competitors into the living rooms of families who would otherwise be deprived of them. The fox, in short, has become the farmer.

The functional tradition

Looking at television from a cultural and semiotic perspective is, however, not the oldest and perhaps not the most respectable academic approach to the medium. That honor belongs to the sociology of mass communications, which grew up largely in the USA where academic interest in television's predecessor in the mass communications field, the press, had always been high.

This tradition has in fact shaped the way most of us think about television, especially in relation to the questions we are pursuing in this essay; namely the relationship between the television message, the everyday reality of the audience, and the functions performed by television for the audience.

We speak here of an academic tradition because, although many of the early theories about television's effect on its audience have been modified, extended or discredited, it remains the case that later research has built upon and not entirely supplanted the assumptions inherent in the early work. For this reason, it is necessary to be aware important of those underlying assumptions. There are at least three, which can be best thought of under the following headings: (1) individualism; (2) abstraction; and (3) functionalism.

1. *Individualism*

This assumption presupposes a one-to-one relationship between the mass communicator and the individual viewer which is justified by reference to the one-to-one model of face-to-face communication. From this assumption, which leaves out of account that much of an individual's response is culturally determined and not internally motivated, has grown the habit of regarding the television viewer as an individual with certain psychological needs. He takes these needs with him to the television screen, and the mass communicator attempts to gratify them. Hence television is seen as a "need-gratification" medium.

2. *Abstraction*

Here the assumption is that an individual's psychological needs are much the same no matter what society or culture he belongs to. Certainly a man's culture can be included as one of the factors which influence these needs, but nevertheless the basic notions imply a kind of universality and timelessness about human relations, which derive no doubt from humanist myths about the existence of a universal "human nature." As a result, this approach tends to disregard the historical processes which have produced such formative developments as the division of labor, class oppositions, regional cultures, economic differentials and the various subcultures, in favor of general psychological needs.

3. *Functionalism*

This approach assumes that television is used by its viewers to satisfy their psychological needs, in a more or less conscious and active way. Functional analysis concentrates on the relations between the different parts in a system, in order to discover how they work and the functions they perform. In respect of television the relationships between the (individual) viewer, the communicator, the channels used and such external factors from the social and cultural experience of the viewer as can be indentified (or better still quantified), are all described in terms of their effect upon each other. The notion of functionalism derives from a well-established sociological discipline, and in the field of mass communication it has extended the range of the earlier stimulus-response assumptions. The most recent research has developed from this into what is called the "uses and gratifications" theory.

Uses and gratifications

We can now refer to some of the specific needs posited by sociologists of mass communication. These needs are always derived ultimately from the individual psyche. Katz *et al.*[2] (1973) list five basic needs to be filled by the mass media. They are:

1. Cognitive needs: the acquiring of information, knowledge and understanding.
2. Affective needs: the need for emotional and aesthetic experience, love and friendship; the desire to see beautiful things.
3. Personal integrative needs: the need for self confidence, stability, status, reassurance.
4. Social integrative needs: the need for strengthening contacts with family, friends and others.
5. Tension-release needs: the need for escape and diversion.

Similarly De Fleur and Ball-Rokeach (1975)[3] propose an "integrated theory" of mass media effects, in which the idea of needs becomes the basis for understanding the media as a whole. Their three categories of need are:

1. The need to understand one's social world.
2. The need to act meaningfully and effectively in that world.
3. The need for fantasy-escape from daily problems and tensions.

These needs supply the basis for De Fleur and Ball-Rokeach's "dependence theory," which is taken to indicate that everyone in modern urban industrialized western society is psychologically dependent to a great extent on the mass media for information which enables them to enter into full participation in society.

One of the most striking examples of the application of this approach came in the study by Peled and Katz (1974)[4] of the media functions in Israel during the war of 1973. Not surprisingly, they found that "dependence" on the mass media tended to increase in a society in crisis. In this situation, individuals' expectations for "information and interpretation" from the media were in fact largely supplied by them. The audience were described as "active" in the sense that they turned to the media with explicit expectations. However, Peled and Katz assert that the "manifest analysis of message content" (into categories like news, entertainment, etc.) is not necessarily sufficient to predict the use to which that content will be put. Especially in the case of television, the messages served not merely the need for information but also the "need for relief from tension and for a feeling of social connectedness."

The hypodermic needle

Clearly the individualist-functionalist view of television does take account of the context within which the television message becomes meaningful. But it also tends to assimilate that context as one of the given variables in a highly predictable set of influences on the individual. Hence any breakdown in the flow of communication from the mass communicator to the individual is often as not seen as a kind of "dysfunction" operating to the detriment of the individual's personal satisfaction. Wright (1964) discusses such dysfunctions in relation to the gratification of the individual's need for surveillance of his society—a need usually seen as being gratified by the provision of quantities of news. But news itself, says Wright, might be dysfunctional: "For example, large amounts of raw news may overwhelm him and lead to personal anxiety, apathy, or other reactions which would interfere with his reception of the items of news about the environment necessary for his normal operations" (p. 107).[5]

Indeed, for Wright, even "handling" (editing) the "large amounts of raw news" might still leave the individual operative in a dysfunctional state, since the news itself "sometimes increases personal tensional tensions and anxiety which, in turn, leads the individual to reduce his attention to the news (hence disturbing the normal state of equilibrium)." For Wright, then, it is appropriate that the bread of news should be leavened by the judicious addition of quantities of entertainment. The function of entertainment, in its turn, is "to provide respite for the individual which, perhaps, permits him to continue to be exposed to the mass-communicated news, interpretation, and prescriptions so necessary for his survival in the modern world" (p. 108).[6]

The language used by Wright is illuminating. He seems to imagine the individual viewer of television as a mechanism whose needs are merely those of equilibrium-maintenance and continued "exposure' to the "necessary" doses of news. Entertainment is merely an anodyne, rather than a positive characteristic of television in its own right. This is in fact a classic statement of the "hypodermic needle" model (sometimes called the "bullet theory") of mass communication's uses and effects. Wright himself has modified this stance in more recent studies (Wright 1975),[7] during the course of which he has given a succinct summary of the hypodermic needle model. It is derived, he notes, from the equally simplistic notion of the audience as a mere "mass." As a mass, an audience displays the following four characteristics:

1. It comprises people from all walks of life.
2. It comprises anonymous individuals.
3. Its members share little experience with each other—there is little interaction between them.
4. Its members are disunited—they are very loosely organized.

Wright comments:

Usually accompanying this concept of mass audience is an image of the communications media as acting directly upon individual audience members—reaching each member or not, influencing him or her directly or not. This view of mass communications has been called the "hypodermic needle model:" each audience member in the mass audience is personally and directly "stuck" by the medium's message. Once the message has stuck someone, it may or may not have the influence, depending on whether or not it is potent enough to "take." (1975, p. 79)[8]

The modification Wright proposes to this simple model is one whereby "a conception of the audience has emerged in which greater notice is taken of the social context within which each audience member operates." But here again the context is seen merely as a variable, rather than a fundamental determinant of the process of communication.

One of the most striking results of the approach we have been describing can be observed in its attitude to individuals who for whatever reason are under-exposed to media influences. They come to be regarded as sufferers from a new kind of social malady, namely "media deprivation." This notion requires that we define the media primarily in terms of their function of disseminating news and public affairs from an informed elite to a dependent mass public. It ignores both the use to which the public might put any given message, and the fact that people who choose not to watch the news are capable of leading relatively normal and fulfilled lives. Hence this approach can be seen as a manifestation of the same ideology which, in the 1960s especially, discovered "verbal deprivation" and "cultural deprivation" in all social groups other than the educated elite. This ideology impinged on the world of practical politics through the attempts that were made to rectify these cultural deficiencies with educational "programs of intervention." (For a good analysis of the "cultural deprivation" debate, see Open University 1972.)[9]

Notice also that this kind of sociological analysis of the mass media's relation with our culture does not seek to go beyond an account of opinions and attitudes of individual people. In this sense television can no doubt have an influence, but it is worth remembering that the opinions and attitudes of most people are chosen from a set which is circumscribed very largely by the modes of thought, the ways of seeing which a culture negotiates for itself. Television cannot easily step outside this central set of choices, and if it did its influence would no doubt be marginal.

Mass consumption or mass communication?

In fact much previous research into the relationship of mass communication with its audience has been more interested in its "massness" than its efficacy as communication. Part of the reason for this has been the inevitable feedback into sociological research of data and assumptions deriving from research done by the broadcasting institutions or their agencies (Neilsen for instance in the USA, JICTAR and BBC Audience Research in the UK).

This kind of research has led to a commonly held view of the audience which has at least colored the findings of numerous academic inquiries over the years; namely the definition of the audience as a "market," comprising people whose shared characteristic is that they are all "consumers" of television output. McQuail (1975)[10] has pointed to the contradiction in this definition:

Audience research, as usually conducted, is a form of market research and hence represents the audience as market—a body of consumers of a particular product. We do not normally regard recipients of communication in other contexts in this way: the people we talk to are not 'consumers' of our words, children are not a market for their lessons, employees of an organization are not consumers of organizational messages, nor are voters a market for the appeals of political leaders. (pp. 187f.)

Naturally, the broadcasting institutions themselves are locked into the prevailing market economy, and it is perhaps tempting for their executives to think in terms of moving a "product" to a "market." But this language is of course metaphorical; audiences do not "buy" television messages, and they do not "consume" the messages transmitted to them.

Hence in order to understand their relationship to the medium, and the effect or otherwise of that medium in their lives, we must concentrate on the dimension which market-oriented research at least most often ignores: the fact that television is *communication* as well as *mass*. As McQuail suggests:

. . . the problem in obtaining or interpreting evidence about effects in general lies partly in the inappropriateness of many formulations of the process of mass communication, a failure to acknowledge that this is a subtle and complex process, a matter of bargaining, interaction and exchange just as much as conversation is between two people. (p. 191)[11]

Clearly an approach which is based on the communicative characteristics of "bargaining, interaction and exchange" (not to mention the negotiation of meaning) will be interested not merely in the number, purchasing power, socio-economic status or opinions and attitudes of the audience. Such an approach must take into account the way in which the members of the audience communicate with each other, and the elements which go to make up an act of communication. Since communication is essentially interpersonal and not intrapersonal, a psychological approach can have only limited relevance to the study of communication. We can recognize that even though an individual's psychological motivation does mediate his behavior, that behavior is normally oriented towards the external world rather than to his internal mental state (see Elliott 1974).[12]

If the mass media are to influence individuals, then, as a form of communication, they must do so according to the rules of communication in general. How does any communication actually influence the receiver of a message? McQuail has isolated five conditions, while stressing that

. . . the central and universal feature of communication influence is the voluntary compliance of the receiver to the sender. The relationship between them is a power relationship . . . The forms of compliance are diverse, and the social situations are usually too complex to find a one-to-one relationship between a type of power and a given case of influence. (p. 163)[13]

The five general conditions which bear upon the effect of communication are:

1. The greater the monopoly of the communication source over the recipient, the greater the change or effect in favor of the source over the recipient.
2. Communication effects are greatest where the message is in line with the existing opinions, beliefs and dispositions of the receiver.
3. Communication can produce the most effective shifts on unfamiliar, lightly felt, peripheral issues, which do not lie at the center of the recipient's value systems.
4. Communication is more likely to be effective where the source is believed to have expertise, high status, objectivity or likeability, but

particularly where the source has power, and can be indentified with.
5. The social context, group or reference group will mediate the communication and influence whether or not it is accepted. (pp. 157-63)[14]

As a result, we can expect the direct effect of *mass* communication upon the individual's behavior and attitudes to be precisely as McQuail suggests: "either non-existent, very small, or beyond measurement by current techniques" (p. 191). Beyond this lame and impotent conclusion the most we can expect of the mass media is that they can provide "frames of reference" and "cognitive detail" about the world: the famous "agenda-setting" effect of the media which can supply for the different audience groups "a quite uniformly held and specific "definition of the situation'" (p. 192).[15] Of course, any definition of the situation which can be supplied under the five conditions noted above and still tend to "override personal experience," must communicate a message that is very close to the culture's collective center, one which people can accept in the knowledge that it derives from deeply held and widely diffused ways of interpreting the world.

If mass communication is indeed communication, then it must communicate something. McQuail *et al.* (1972)[16] have classified under four headings the relationship of media content (what it says) to audience use (what we do with it). They suggest that this relationship has to do with:

1. Diversion
   (a)escape from the constraints of routine
   (b)escape from the burdens of problems
   (c)emotional release
2. Personal relationships
   (a)companionship
   (b)social utility
3. Personal identity
   (a)personal reference
   (b)reality exploration
   (c)value reinforcement
4. Surveillance—maintaining an overall view of the immediate environment. (p. 155)[17]

It is interesting that this classification was erected partly in response to the inadequacy of previous theories, which had overstressed the escapist function of television. And yet the idea of television communication providing escape is still firmly embedded in the Diversion category.

But escape is not just a matter of the inhabitants of Jubilee Close watching *Coronation Street,* though that kind of self-reflexive escape is indeed popular. It is important to remember that the "bargaining, interaction and exchange" of the communication process *itself* functions as part of the entertainment, bonding us as viewers, via the message, to the reality of our culture, and thus lifting the burden of an isolating individualism from our shoulders. This is not the kind of gratification, however, that is easy to articulate for social scientists' questionnaires. It's easier to say that "TV takes you out of yourself."

Who is the communicator?

If the relationship of the audience to the communicated message is a peculiarly complex one, then so is the relationship of the audience to the communicator. To begin with, analysts of mass communication are faced with an apparent paradox which does not occur in most other forms of communication. Neither party in the communication act knows who the other is. Normally when one party encodes a message according to one set of codes and conventions and a second party decodes that message according to different codes, the result is known as *aberrant decoding.* For example, the masons and the glaziers who built Chartres Cathedral constructed their statues and windows to communicate to members of their well-defined and relatively cohesive culture their sense of man's place in relation to time, God and the revelations of the testaments, by means of certain familiar sacred stories told in stone and glass. The modern perceiver of these same messages may well decode them as perfect examples of gothic art, or as manifestations of the medieval delight in resemblances, numerology, etc. The modern decoding is *aberrant* in relation to the communicative intentions of the encoder. The cultural context of the two parties is different.

As Eco (1972, pp. 104-6)[18] has pointed out, aberrant decoding is quite normal in the field of the mass media. Logically this must be the case since the professional encoders and the "undifferentiated mass of receivers," respectively, cannot be certain in a complex society like ours that they speak the same language (see Smith 1973, pp. 49-50).[19] However, this very characteristic of the television communication imposes a discipline on the encoders that ensures that their messages are in touch with the central meaning systems of the culture, and that the codes in which the message is transmitted are widely available. The very possibility that the audience might use the message in ways not intended by the encoder has a double effect: it drives the professionals towards the center of their culture (just as the masons and glaziers of Chartres used their plastic and visual media to communicate messages already very familiar in their culture); and, conversely, it protects the audience members from direct coercion or even influence towards any particular version of the truth.

A further complexity we face in understanding the communicator's relation to the audience is the problem of identifying just who "he" is. In fact we can identify three simultaneous levels in the presentation of the communicator. They are:

1. The image on the screen, whether that image be one of a newscaster, the actors in a fictional representation, or camera shots of the world out-there—urban streetscapes, for instance.
2. The broadcasting institution, its employees and professional codes (which Smith 1973, p. 5,[20] has called the "ideology" of "objectivity" or "impartiality").
3. The culture for which the messages are meaningful. At this level the communicator is the macro-group, of many millions of people, of which each member of the audience group is a differentiated part.

Hence we can say that culture communicates *with itself* via the mediation of the (second level) professional communicators who are manifested as the (first level) encoded messages on the screen. It follows that the television medium is not a closed system, obeying its own internal rules and relatively uninfluenced by "external" conditions. As Hall (1973)[21] has pointed out, the professional communicators themselves are not insulated from cultural influences. The people who are responsible for what we call "production" of output mediate the messages but do not *originate* them. They "draw topics, treatments, agendas, events, personnel, images of the audience, 'definitions of the situation' from the wider socio-cultural and political system, and so they cannot constitute a 'closed system'"(p. 3).[22]

Thus we can say that there is no single "authorial" identity for the television communicator. Futhermore, the image on the screen would hardly be able to make itself understood at all were it unable to rely upon the resources of everyday verbal language. After all, most heads on television are "talking heads." For this reason it is perhaps instructive to remind ourselves that television's functions are to some extent dependent upon and defined by the functions performed by speech in general. Of course, television is a semiotic system going beyond mere words—but much of its visual content takes the form of "paralinguistic" signs derived ultimately from pre-televisual (real-life) linguistic codes. We have argued in this essay that television functions in society as a form of communication. But the language upon which its codes are modeled itself performs several discrete functions. The linguist Jakobson (1958)[22] has isolated six of them, all of which can be observed at work on television; indeed many of its messages seem to serve little purpose other than to perform them. Briefly, they are as follows:

1. *The referential function*. Language's most familiar function, where the relationship between a sign and its *referent* or object is dominant.
2. *The emotive function*. This concerns the relationship between a sign and its *encoder*; it expresses his attitude towards the subject of the message (e.g. "it's been a long day" communicates the speaker's attitude towards the day, and does not refer to its time-span).
3. *The conative function*. This concerns the relationship between the sign and its *decoder*. Imperative commands are messages where the cognative function is most clearly dominant.
4. *The poetic function*. Here the dominant function is the message's concern with *itself*. It is not confined to poetry. Slogans can use it (e.g. "I like Ike"), and so can proverbial sayings (e.g. "Finders, keepers, losers, weepers"). It is dominant in many television advertisements.
5. *The phatic function*. Here the dominant function of the message is to stress the *act of communication* between the parties involved. Remarks about the weather are its classic example. Television programs use it frequently; perhaps because of, rather than in spite of, the "remote" relationship between broadcaster and viewer.
6. *The metalinguistic function*. Here the function of language is to communicate a message *about language*. Literary criticism is all metalanguage, and many television programs, especially comedies, are "about" the television message. Parodies like *Monty Python, Rutland Weekland Television*, etc., are especially adept at exploiting metalanguage.

Notes

[1] Rapoport, A. (1969) 'A system-theoretic view of content analysis', in G. Gerbner, O. Holsti, K. Krippendorf, W. Paisley, and P. Stone (eds), *The Analysis of Communication Content*. New York: John Wiley, pp. 123-32.

[2] Katz, E. Gurevitch, M. and Hass, E. (1973) 'On the uses of mass media for important things', *American Sociological Review*, Vol. 38, pp. 164-81.

[3] DeFleur, M. and Ball-Rokeach, S. (1975) *Theories of Mass Communication*. New York: McKay.

[4] Peled, T.and Katz, E. (1974) 'Media functions in wartime: the Israeli homefront in October 1973', in J. Blumler and E. Katz (eds), *The Uses of Mass Communications*. Beverly Hills, Ca.: Sage, pp. 49-69.

[5] Wright, C.R. (1964) 'Functional Analysis and mass communication', in L. Dexter and D.M. White (eds), *People, Society and Mass Communications*. Glencoe, Ill.: Free Press, pp. 91-109.

[6] Ibid.

[7] Wright, C.R. (1975) *Mass Communications: A Sociological Approach (2nd edn)*. New York: Random House.

[8] Ibid.

[9] Open University (1972) *Sorting them out: two essays on social differentiation*. Milton Keynes: The Open University, (E 282 Unit 10), pp. 55-140.

[10] McQuail, D. (1975) *Communications*. London: Longmans.

[11] Ibid.

[12] Elliot, P. (1974) 'Uses and Gratifications Research: A Critique and Sociological Alternative', in J. Blumler and E. Katz (eds), *The Uses of Mass Communications*. Beverly Hills, Ca.: Sage, pp. 249-86.

[13] Ibid.

[14] Ibid.

[15] Ibid.

[16] McQuail, D., Blumer, J. and Brown, J. (1972)'The television audience: a revised perspective', in McQuail (ed.) *op. cit.*, pp. 135-65.

[17] Ibid.

[18] Eco, V. (1972) "Towards a Semiotic Inquiry into the TV Message", *WPCS*, No. 3, pp. 103-21.

[19] Smith, A. (1973) *the Shadow in the Cave*. London: George Allen Unwin.

[20] Ibid.

[21] Hall, S. (1973) 'Encoding and decoding in the television discourse', Centre for Contemporary Cultural Studies, Birmingham, *Occasional Papers*, No. 7.

[22] Ibid.

[23] Jakobson, R. (1958) 'Closing Statements: Linguistics and Poetics', in T.A. Sebeok (ed.) (1960) *Style and Language*. Cambridge, Ma: M.I.T. Press, pp. 350-77.

TELEVISION CULTURE

Hal Himmelstein

TELEVISION is frequently referred to as our most "popular art." Indeed it is hard to argue against television's pervasiveness in our culture—almost every household in the United States and a great majority of households in other industrialized societies have at least one working television receiver, and current data suggests that the average American household's television set is operating nearly seven hours a day or almost half of our waking hours. But the term "popular art" has come to mean much more than television's pervasiveness or seeming preoccupation with staring at the electronic cyclops. "Popular art" has taken on some rather ominous overtones as something lesser in quality than those traditional cultural products termed, in contrast, "high art." Television is even discussed in many circles as "subart" or "nonart" in comparison with the traditional creative products of literature, drama, music, visual arts, dance, and yes, even film. Such a popular art/high art dichotomy, is in my view, much too easily arrived at and neither a reasonable nor a responsible approach to evaluating the creative products of television's creative community. Let's examine this area more carefully.

The mass culture-high culture debate

From the 1930's through the 1950's a debate raged among academicians and literati over the nature and meaning of what had come to be called "mass culture" or "popular culture." Included in the mass culture category were television and other so-called "public" art forms (radio, film and a variety of popular literature and popular music forms). While mass culture was never precisely defined, it seemed to refer to certain objects and events that engaged the attention of all strata of the population as opposed to those objects and events that were traditionally assumed to be the private domain of some ill-defined cultural elite. These latter objects and events were called high culture or superior culture to distinguish them from mass culture.

The most articulate participants in this debate over the nature and sociocultural significance of mass culture were Dwight Macdonald, Edward Shils, and C. Wright Mills.

Macdonald, who drew upon the seminal writings of art critic Clement Greenberg, advanced the argument that "Mass Culture began as, and to some extent still is a parasitic, a cancerous growth on High culture."[1] Macdonald was forced to admit that even in the "old" art forms such as literature, music, painting, dance, and architecture there could be found numerous examples of mass culture. But he then made a critical distinction between these art forms and the more recently emergent art forms: *Mass Culture has also developed new media of its own, into which the serious artist rarely ventures: radio, the movies, comic books, detective stories, science fiction, television.*[2]

According to Macdonald's perspective, high culture has traditionally been the province of the serious artist "communicating his individual vision to other individuals"—an audience of intellectual and wealthy patrons, a cultural elite. Mass culture, in contrast, is the province of the individual artisan or team of artisans engaging in the "impersonal manufacture of an impersonal commodity for the masses."

To Macdonald, media such as television, film, radio, comic books, detective stories, and science fiction were initiated through a technological imperative that made possible the inexpensive reproduction and distribution of *standardized artistic products* in great quantities for consumption by a large group of persons who had become homogenized through the process of democratization.

Macdonald concluded that mass culture is not only reinforcing the plebian tastes of the masses, but is wrecking the sacred province of the cultural elite who are dwindling under the pressure of massification as the intellectuals who comprised the once-strong elite are displaced by highly trained specialists who are themselves caught up in the mass production cycle.

Two weaknesses in Macdonald's basic arguments should be noted. First, Macdonald set up a "straw concept"—mass culture—which he then knocked down on the basis of the presumed level of taste of an audience, the "masses." Such an approach flies against the notion that each work of art should be judged on its own merits as to both its formal structure and its handling of the symbol systems or cultural myths internal to the work and reflecting back on the culture in which the work was created. Instead, Macdonald produced generalized statements about art forms without citing representative works in those art forms that might contradict the very notion of the "mass-ness" of the art forms themselves.

Second, Macdonald greatly overgeneralized when he argued that serious artists rarely ventured into his so-called mass culture "entertainment." Macdonald chose to ignore a number of film artists who were initially recognized for their work in other art forms before they became filmmakers and who as filmmakers made significant contributions to the enhancement of the filmic art. Examples of such film artists include the Russian filmmaker Sergei Eisenstein, who began work as a director of Soviet theater and later became a noted filmmaker and film theoretician; Jean Cocteau, who began as a theater actor, and was a successful poet, novelist, dramatist, and essayist before becoming a highly respected filmmaker; and, more recently, American experimental filmmaker Jordan Belson, who was a widely exhibited painter prior to his extensive involvement with film.

In an apparent attempt to shore up the crumbling walls protecting what he called a cultural elite, Macdonald failed to properly assess culture, namely, experience cast up by an individual or individuals in symbolic forms that are capable of being apprehended by other individuals. Such symbolic forms may heighten our awareness of continuities or discontinuities in our personal environments and our broader cultural contexts; they may also entertain us, providing us with diversion from the harsher everyday "realities" of living in the world while at the same time establishing their own set of realities by their very presence in the world. Heightened awareness and entertainment are, in this cultural context, not mutually exclusive.

Edward Shils, like Macdonald, noted that in Western societies today there has occurred a loosening of the power of tradition. Shils interpreted this to mean that individuals in these societies have a greater variety of choices of personal experience than in any other period in recorded human history. Shils wrote that the choices today are more personal, and not dictated as much by tradition, scarcity, or authority as in the past. Shils noted the virtues of this "mass society": *The new society is a mass society precisely in the sense that the mass of the population, has become incorporated into society. . . most of the population (the "mass") now stands in a closer relationship to the center than has been the case in either premodern societies or in the earlier phases of modern society.*[3]

As the masses become integrated into the society's center—its central institutions and the value systems that legitimize those institutions—their perception of cultural events is heightened.

Thus Shils granted what Macdonald most feared—that the cultural elite (or what Shils terms "the classes consuming culture") may well become fewer in number and less discriminating in their preference for certain works of art. However, Shils, using an argument that attached importance to greater numbers (i.e., the "masses") as a means for measuring the significance of taste in a society, rejected Macdonald's notion that art is doomed through the growth of mass culture. Thus, a basic and very important distinction between the two arguments becomes apparent: Macdonald argued that mass culture will continually decline in taste level as fewer products of high culture become available for imitation, while Shils argued that mass culture will continue to rise in taste level as more individual's consciousness levels are raised by their exposure to art forms of which they had previously been unaware.

Shils devised a tentative three-category paradigm of culture levels, which he wrote were "levels of quality measured by aesthetic, intellectual, and moral standards."[4] The levels were differentiated as superior or refined culture, mediocre culture, and brutal culture. Superior culture was characterized according to the seriousness of its subject matter based upon some standards of truth and beauty. This culture, Shils suggested, was produced by a "high intelligentsia" with international ties and was consumed primarily by intellectuals (e.g., university teachers and students, members of learned professions, artists, writers, journalists, and higher civil servants). Mediocre culture was characterized as not measuring up to the standards of superior culture—it was less original culture. This culture, according to Shils, was produced by a "mediocre intelligentsia" and was consumed by the "middle class." Brutal culture was characterized as being more elementary in its symbolic elaboration (i.e., containing more directly expressive actions with a minimum of symbolic content). It lacked the subtlety and depth of penetration of either superior or mediocre culture. It continued traditional cultural patterns with little consciousness of their traditionality. Brutal culture was produced by what Shils termed the "brutal intelligentsia" who had no connections with superior culture and was consumed by the members of the industrial working class and the rural population. Shils added that there was an intermixture of the three cultures in the newspaper, on television, and in film, although one would most likely find mediocre culture pervading them.

It must be clearly recognized that Shils was himself leery of claiming that his culture categories were rigid. For example, he was quick to note that even in the brutal culture category there would be an occasional individual work that reached the limits of superior culture. In fact, in a perceptive footnote to his discussion, Shils made a frank admission as to the limitations of his culture paradigm:

I have reservations about the use of the term "mass culture," because it refers simultaneously to the substantive and qualitative properties of the culture, to the social status of its consumers, and to the media by which it is transmitted. Because of this at least three-fold reference, it tends to beg some important questions regarding the relations among the three variables. For example, the current conception of "mass culture" does not allow for the fact that in most countries, and not just at the present, very large sections of the elite consume primarily mediocre and brutal culture. It also begs the important questions whether the mass media can transmit works of superior culture, or whether the genres developed by the new mass media can become the occasions of creativity and therewith a part of superior culture. Also, it does not consider the obvious fact that much of what is produced in the genres of superior culture is extremely mediocre in quality.[5]

Thus, while Shils was arguing in support of the effects of mass culture in raising the aesthetic, social, and moral consciousness of the so-called masses to some higher level than they had previously enjoyed, he was not foolish enough in his polemic to establish rigid boundaries of culture classification that would preclude consideration of individual works of art as something other than representative of certain classes of culture. Nevertheless, his argument is not sufficient even given this limited flexibility—for by and large he was still matching genres with culture classes, means of transmission, and presumed taste levels of audiences.

C. Wright Mills concentrated his criticism of mass media and mass culture on the notion that the media themselves are the tools used by the modern elite in their efforts to manipulate the masses.

According to Mills, the result of this elite manipulation of the masses is the impersonalization of life itself:

As they now generally prevail, the mass media, especially television, often encroach upon the small-scale discussion, and destroy the chance for the reasonable and leisurely and human interchange of opinion. They are an important cause of the destruction of privacy in its full human meaning. That is an important reason why they not only fail as an educational force, but are a malign force: they do not articulate for the viewer or listener the broader sources of his private tensions and anxieties, his inarticulate resentments and half-formed hopes. They neither enable the individual to transcend his narrow milieu nor clarify its private meaning.[6]

Mills here provides some insight into what works of art might be expected to provide an audience, namely, an articulation of the broader sources of the audience member's "private tensions and anxieties, his inarticulate resentments and half-formed hopes." As the audience member begins to apprehend the work's significations, he should be able to transcend his private world and become aware of a broader cultural milieu as it relates to his own environment. Thus works of art can serve a culturally integrative function in the sense of promoting better understanding of our culture.

The difficulty with Mills' statements about mass culture (like those of Macdonald and Shils) is located in his sociological overgeneralizations. By producing a blanket indictment of "mass media" as tools for cultural manipulation by an elite, Mills refused to grant the possibility that artists are producing culturally significant "liberating" works of art in such media. Those same works of art that promote cultural understanding may at the same time act as a freeing influence—one may better understand the culture in which he lives and thereby reject elements of that culture as not meeting his personal needs.

Shils at least recognized the need to consider the individual work of art and its apprehension by a spectator in some larger cultural context. Yet he, like Macdonald and Mills, was never able to get to the heart of the issue. The major limitation of the mass culture/high culture debate is that it tells us little or nothing of the aesthetic designs of works of art as representative examples of genres or of their interpreted symbolic or mythic significance in the culture in which they are produced. Rather, the debate assumes that some highly ambiguous changes in individuals' lives are brought about through their exposure to certain art forms; these changes are described as "modifications of levels of taste" and "manipulation of one group of people by another." These limiting notions divert our attention from the central problem, which is to uncover how works of art are integrated into people's understanding of the symbolic forms or "myths" that comprise the shared experience of which the works are a significant part.

Unlike the sociologists Macdonald, Shils, and Mills, aesthetician Abraham Kaplan has provided us with some specific criteria that we might apply to individual works of art to determine into which area of the high art/popular art dichotomy the works might fall. While Kaplan's criteria are on the whole plausible, it should be recognized at the outset that they, as in the case of any paradigm, are subject to modification as new directions in art creation are explored.

Kaplan distinguished art (or high art) from popular art by noting that popular art may be said to be "mass art"—art that is "mass-produced or reproduced, and is responded to by vast numbers of people."[7] However, Kaplan quickly qualified this definition by noting that the specification of the origin and destination of works of art

. . .does not of itself determine just what it is that is being responded to. There is no fixed a priori relation between quantity and quality, and especially not between quantity and certain specific qualities as distinguished from worth in general.[8]

Kaplan believed that popular art forms have their own formal structures and meanings, and as such are not bastardizations of high art or the products of external social forces. To support his contention, Kaplan divided the elements distinguishing art from popular art into three general classes: form in the work of art itself, the relationship between the work of art and the spectator confronting the work, and the social functions of the work of art. A schematic representation of Kaplan's high art/popular art criteria follows.

FORM

POPULAR ART	HIGH ART
1. Provides a form with cultural significations.	1. Provides a formula, but gives us nothing to which to apply it.
2. All elements in the work are significant.	2. A few elements in the work are singled out as carriers of meaning, while the remaining elements are merged into an anonymous mass.
3. Fill-in by the spectator is called for.	3. The work is predigested (i.e., self-completed).
4. The work is a cognitive challenge to the spectator due to its *ambiguity*.	4. The work is intolerant of ambiguity.
5. The work is illusory without being deceptive.	5. The work rings false.
6. In the work, reality is transformed from represented subject to expressive substance.	6. The work takes over the shapes of reality, but not the form.

RELATIONSHIP BETWEEN THE WORK OF ART AND THE SPECTATOR

POPULAR ART	HIGH ART
7. Emotion or feeling is expressed by the work.	7. Emotion or feeling is associated with the work.
8. Feeling is drawn out of ourselves by the work, creating a fulfilling aesthetic experience.	8. Feeling is consummated in the work, but without the spectator's aesthetic fulfillment.
9. The work creates an intense awareness (i.e., sensibility) in the spectator.	9. The work creates superficial feelings (i.e., sentimentality) in the spectator.
10. The work enlarges the spectator's apprehension of a "world."	10. There is too little of substance that requires a spectator's understanding of the work and its cultural significations.
11. The work produces spectator empathy or identity with its substance.	11. The work produces spectator self-centeredness.
12. The work helps the spectator transform ugliness in the work into meaning.	12. The work helps the spectator escape ugliness.
13. The work dares to disturb the spectator.	13. The work seeks to relieve the spectator's anxiety.
14. The work heightens our aesthetic perception or response so that we become self-stimulated when faced with the work itself.	14. The work produces our recognition of its substance, but this is a mere reaction to the work.
15. The work raises us to "divine objectivity."	15. The work is too subjectively human.
16. The work presents to immediate experience the values of a culture in factual form.	16. The work assures us that the facts as presented support our prejudged values.
17. The work is wish-fulfilling in itself.	17. The world depicted in the work is wish-fulfilling.
18. The work transforms reality only to enable the spectator to better apprehend it.	18. The work strips fantasy of the qualities of creative imagination.

19. The work posits a vision of reality and at the same time evokes a pleasurable experience in the spectator.	19. The work evokes a pleasurable experience only.
20. The work calls for enough aesthetic distance by the spectator to give him perspective and enough wisdom to enable him to see himself in perspective (i.e., objectification).	20. The work calls for little or no aesthetic distance.
21. The works shows us the limits of our powers in the world.	21. The work turns its back on a world it has never known.

SOCIAL FUNCTIONS OF THE WORK OF ART

22. The work appeals to a common denominator—"the universality of art."	22. The work appeals to a "distinctive majority taste" and is indicative of the status quo in culture (i.e., imitation of previously successful forms).

Kaplan's criteria offer us one approach that differentiates works of art according to certain specific characteristics of the works themselves. However, Kaplan's high art and popular art categories are so meticulously constructed that they preclude the probability of any single work of art satisfying all the prerequisites he established for "high art"; on the other hand, many works of art, including some of those commonly classified as "popular art," could seemingly satisfy many if not most of those prerequisites. Despite such limitations, Kaplan's approach is valuable as a reference against which one can compare evaluations of works of television art.

Since one major concern of this essay is the examination and evaluation of television and video as culturally significant art forms, one of Kaplan's conclusions about his high art/popular art dichotomy is relevant here:

For audiences, art is more of a status symbol than ever; its appearance in the mass media is marked by a flourish of trumpets, as befits its status; the sponsor may even go so far as to omit his commercials. I am saying that even where popular art vulgarizes yesterday's art, it might anticipate tomorrow's—baroque once meant something like kitsch. I am willing to prophesy that even television has art in its future.[9]

Analyzing TV Culture

There is no reason for us to doubt that from the countless hours of television to come, a few significant works of television "art" will emerge to assume their deserved places in the history of this art form. For many culture critics, however, preoccupation with separating television programs into Kaplan's categories is at best tangential to what they perceive to be the central questions of television's "expressive substance." These critics increasingly focus on the nature of the special "worlds" presented on the small screen.

Recent analytical strategies of merit have focused on television as "dream"; on television's priviledged position as reciter of our culture's significant myths; and on television narrative codes as easily comprehended social constructions offering viewers continual reassurance that existing social relations are appropriate.

The notion of television as dream was advanced by history professor Peter H. Wood, who hypothesized that television constitutes an important part of the dream life of our culture. Both television and dreams, wrote Wood, are highly visual; both are highly symbolic; both involve a high degree of wish fulfillment (fantasies or hallucinatory experience), both appear to contain a large amount of disjointed and trivial material; both contain powerful content which is quickly and easily forgotten; and both consistently use materials drawn from recent experience.

According to Wood, television may be linked to a "collective subconscious" that is manifested both in the creative community and in viewers—"a TV society purposefully and unconsciously creates its own video world and then reacts to it."[10] This dreamlife is actualized through dramatization that is often condensed with little or no prior exposition, and contains a variety of powerful verbal and visual symbols pointing to a subconscious world of strong fears and desires.

Wood's linking the individual's dream life with the collective subconscious as actualized through dramatization is an argument for the existence of powerful myths or stories that are played back to us endlessly on television. Psychoanalyst Erich Fromm has pointed to significant connections between the individual's dream life and the culture's myth life. According to Fromm, both the dream and the myth are highly creative acts; both employ highly-charged symbolic language; and both are grounded in the reality of lived experience.[11] While a significant personal experience of the preceding day often evokes a dream, myths recall a quintessential event or state of being in the culture's "timeless past," which constitutes a "sacred" inviolable history always present and intelligible in the myth's contemporary form. Recurrent dreams and myths express a lietmotif of significant main theme—the former in one's life and the latter in the collective life of the culture—which frames the way we tend to view our world.

Both Wood's and Fromm's work moves beyond questions of "artistic worth" and focuses instead on the social relevance of narrative. In an essay entitled "Television Melodrama," professor of English, David Thorburn advanced this discussion to another level by positing that one narrative form in particular—television melodrama—functions as a powerful, socially conservative agent binding the television culture to its past—a past increasingly defined through television itself. Thorburn noted that television melodrama's consistent references to its own history, its repeatedly reassuring conclusions, and its moral allegories presented in topical, familiar contexts are, for the audience, "the *enabling conditions*" for an encounter with forbidden or deeply disturbing materials."[12] For Thorburn, melodrama (which includes the television genres of made-for-television movies, soap opera, fictional lawyer and doctor programs, Westerns, mysteries, and adventure) is above all a market commodity employing stereotyped formulas and expounding "the conventional wisdom, the lies and fantasies, and the muddled ambivalent values of our bourgeois industrial culture."[13] Yet it is precisely these characteristics of melodrama that Thorburn feels highlight its importance as a cultural artifact. The complex reality of contemporary social relations is deflected by the melodramatic aesthetic devices of formulaic plot structuring and condensation that eliminate the need for long establishing sequences providing social context for conflicts and focus attention instead on a few moments of heightened moral conflict. Melodrama provides us with models of clear resolution for excessively enacted conflict. According to dramatist Peter Brooks, melodrama demands strong justice, "a perfect justice of punishment and reward, explusion and recognition," unlike tragedy, which often includes the ambivalence of mercy in its code. Brooks elaborated on the melodramatic aesthetic at work:

Melodrama regularly simulates the experience of nightmare, where virtue, representative of the ego, lies supine, helpless, while menace plays out its occult designs. The end of the nightmare is an awakening brought about by confrontation and expulsion of the villain, the person in whom the evil is seen to be concentrated, and a reaffirmation of the society of "decent people."

In *Ways of Seeing*, critic John Berger argues that "the art of any period tends to serve the ideological interest of the ruling class." In advanced capitalist society, popular art serves such a purpose by forcing the majority of people—its large audience—to "define their own interests as narrowly as possible."[15] *Melodrama's basic aesthetic of reassurance enables the audience member to easily apprehend what appear to be the* central value conflicts of the society in which melodrama is produced—conflicts ultimately defined and resolved by dominant culture producers and program managers.

These approaches to the evaluation of the worlds presented to us through television move beyond a preoccupation with popular culture "weaknesses" as conceived within traditional aesthetic parameters and instead focus on the social contexts in which works of popular culture are produced and consumed.

Any discussion of the social contexts of television production and consumption would be incomplete without including alternative television or artists' video. Artists' video can be distinguished from television in part by its adherence to art's traditional concept of the personal expression, of the individual artist's (or small group of artists') struggle against the ordering principles established by the culture's dominant institutions. As Marcuse noted, "art remains the one place where actual subversion can still take place."[16]

Those who create video works are often the first to claim that television is a dominant cultural institution that does not allow those with individual "vision" access to its high technology production mechanisms—a statement not without merit. They are also quick to condemn contemporary television as mindless pap, a natural outgrowth of a production and distribution system that is governed not by the creative community itself but rather unsympathetic administrators, businessmen, and lawyers who are constantly wary of regulatory scuffles ensuing from the production of controversial material.

Video works produced in response to these guiding concepts frequently take the form of parodying television content, or of demonstrating how on an extremely low budget the artists can one-up television by doing something worthwhile, or of showing their disdain for commercial television's use of the advertisement's time signature—a "linear succession of logically independent units of nearly equal duration," the video of "boredom."[17]

Video often uses deliberate repetition, sexual explicitness, and occasionally candor on social issues to break the taboos of television. When done well, it can be highly effective as an alternative, subcultural voice. In the case of community video documentary such as that produced by Downtown Community Television (Jon Alpert and Keiko Tsuno) and at numerous public access cable channels throughout the United States, the use of relatively inexpensive video technology by individuals sensitive to their community's problems and needs can and has acted as a unifying force within the community—the portapak picks up where the front porch left off, providing an arena for discussion of neighborhood interests.

Video as employed in the art community and in the social community has been evaluated according to traditional aesthetic conceptions of worth and has often been found to be wanting. However, as is the case with television, perhaps different perspectives need to be employed in the evaluation of video works, especially as the works set themselves against the institutional domination of the television *industry*.

Notes

[1] Dwight Macdonald, "A Theory of Mass Culture," in *Mass Culture: The Popular Arts in America*,eds. Bernard Rosenberg and David Manning White.(Glencoe, Ill: Free Press, 1957), p. 59.

[2] Ibid.

[3] Edward Shils, "Mass Society and Its Culture," *Daedalus 89*(Spring 1960): 288.

[4] Ibid.

[5] Ibid.

[6] C. Wright Mills, *The Power Elite*(New York: Oxford University Press, 1956), pp. 314.

[7] Abraham Kaplan, "The Aesthetics of Popular Arts"*The Journal of Aesthetics and Art Criticism 24*(Spring 1966): 352.

[8] Ibid.

[9] Ibid.

[10]Peter S. Wood, "Television as Dream," in *Television as a Cultural Force, Eds.* Richard Adler and Douglass Cater (New York: Praeger, 1976), pp. 17-35.

[11]Erich Fromm, *The Forgotten Language: An Introduction to the Understanding of Dreams, Fairy Tales and Myths* (New York: Rinehart & Co., 1951), pp. 156, 192ff. For a discussion of television and myth, see Hal Himmelstein, *Television Myth and the American Mind* (New York: Praeger Publishers, 1984).

[12] David Thorburn, "Television Melodrama," in *Television as a Cultural Force*, eds. Richard Adler and Douglass Cater (New York: Praeger, 1976), pp. 77-94.

[13] Ibid.,p.80

[14] Peter Brooks, *The Melodramatic Imagination* (New Haven: Yale University Press, 1976), p. 204.

[15] John Berger, *Ways of Seeing* (London: British Broadcasting Corporation, 1972), pp. 86, 154.

[16] Herbert Marcuse, *The Aesthetic Dimension* (Boston: Beacon Press, 1977), p. 69.

[17] David Antin, "Video: The Distinctive Features of the Medium," in *Video Art* (University of Pennsylvania, Institute of Contemporary Art, 1975), p. 67.

WATCHING TELEVISION

John G. Hanhardt

WHAT DOES IT MEAN to "watch television"? In beginning to suggest an answer to this question I will consider what factors control broadcast television, review the content and structure of television programming, and finally examine where and how television is viewed. My frame of reference will be broadcast television in the United States. It was during the years following the Second World War and extending through the 1970's, the period of the rapid economic expansion of late Capitalism,[1] that broadcast television assumed such a powerful and pervasive presence in all social sectors and geographic areas of this country.

In what follows I will not attempt a history of broadcast television but rather outline some of the distinctive features which mark its emergence in our culture and the attitudes the viewer/consumer has taken to its presence. I have linked the terms viewer and consumer because it is the equating of these two attitudes that ultimately defines how we have come to watch television. It is my premise that the viewer's perception of the medium is largely determined by the role television has come to play as an entertainment and information industry shaped by the marketplace of corporate capitalism. The goal of my argument that we must redefine our expectations of what television can become in order for its future to be something other than the repackaging of television's content along the ideological lines of the new global technologies of information processing and distribution. In beginning this reexamination of television we should acknowledge the political activist and artist's use of video since the 1960s in single-channel and installation formats, which have provided the most radical and searching models of what television can be. It will be my contention here that it is essential that we not lose sight of what artists have accomplished when we consider the future of television.

The emergence of broadcast television, its rapid expansion and pervasive place in our society, has been effectively controlled by a monopoly of networks. It is an industry which developed within the guidelines established by a federal telecommunications policy and its laws, which implicity sought to control the distribution and licensing of what was a potential communications technology.[2] The possibility for a proliferation of local networks providing a powerful tool for community expression and communication of alternative ideologies and information was effectively checked. The strategy shaped earlier by the example of radio served as a model for the decisions leading to the introduction of television to the public as a communications technology. In fact it did not function in that way.[3] The capitalization required for television emerged through the development and control of the manufacturers of the technology interlocked with the establishment of the major corporate networks. The institution of television, like radio, was to be fashioned as an Ideological State Apparatus[4] that effectively spoke one voice and served as the means to extend further the marketplace and the ideology of corporate capitalism.

Whereas television was not controlled as a state system, in the sense that a centralized authority approved and passed on its content and policies, there was in effect a control of the latitude of what was permitted to be shown on television. The notion of the middle ground and the myth of balanced journalism together neutralized the medium and transformed its potential for analysis and independence into a one-way street of communication. It was not modeled on a two-way communications system but was rather a conduit for programming. Thus, the very definition of television as a "communications" technology comes up against the fact that the viewer is a passive receiver, unable to communicate, interact and fashion programming. As with the radio listener, the television viewer has few options within a spectrum of programs either produced by the networks, the Public Broadcasting System, or independent stations whose product is limited to what is available in syndication from the networks or is modeled on the dominant program formats of commercial televison.

What is available to the viewer has a profound impact on how he/she watches television. The standardization of programming is on one level, as already outlined, controlled by the monopolization of the major broadcasters and their hegemony over independent stations by virtue of the ideological control effected by the laws together with the commercial standards they articulate and support. On another level television's content is a marketplace for the selling of goods and ultimately a way of life. The "consciousness industry"[5] of broadcast television sells an attitude that is reflected and reinforced in the viewer's relation to the television program itself. In other words, the broadcast industry has commercialized the airwaves. The advertisements which punctuate the broadcast time are of a piece with the structure of the broadcast day. In order for this technology, for the television set, to effectively work as a selling tool it must function within a context that induces an attentiveness that will effectively be absorbed by the largest number of viewers/consumers.

The television set, as opposed to the experience of going to the movies, does not operate in theatrical situations. It again followed upon the success of the radio receiver in being marketed for the home environment which includes not only the permanent living space of the individual and family but such neutral territories as the hotel/motel room. With few exceptions, such as taverns where television provides background entertainment or sports, television has become a focal point for the consumption of entertainment, news, and new products for the home in which it is being viewed. The television set in the home is a focal point replacing the fireplace which in earlier times marked the time of day in its rhythm and was the place where the society/culture's public consciousness and myths were transmitted by reading and speech.

The television set primarily functions within the living room/common space and the bedroom of the house/apartment. Programming is structured around thirty and sixty minute time blocks which are subdivided by the regular introduction of commercials. This temporal framework establishes an expectation on the viewer's part, a narrative flow and rhythm, which is designed to showcase the commercial. Entertainment, news, information, and commercials are treated in commerical television as building blocks to be formed into shorter temporal units that are then shaped into a seamless narrative and world view. Television is telling stories, fictions about products, life styles, historical and contemporary events that deliver the myth of television's being a window onto the world.

The technology of television reinforces the notion of the medium providing immediate access to what we see on the screen. It is a technology which is essentially mysterious and unavailable to the home viewer. The television set is a closed container to which only trained specialists can gain access. In order to watch television all one has to do is plug it in. Television, unlike the motion picture which we know to be recorded on film,[6] has built within it since its earliest days the mythology of the live broadcast. Implicity within our perception of television is that it is immediately delivering events to us. The soundtrack of the studio audience, the mistakes of the performers, the sports events, and the urgency of news events appear live but are in fact, for the most part, all effects deliberately added, edited, and manipulated to convey an urgency and feeling of being live.

The view of television as a live broadcast medium beaming into our living spaces the events of the day and entertainments of popular culture is reinforced by how we perceive the television set and the environment in which it is situated. The small size of the television screen housed within a console or portable model is essentially a piece of furniture. The receiver is an appendage of the appliances and furniture we are accustomed to using in our homes. Its programs, especially the commercials, are prepared in order to be heard on the soundtrack, as well as seen, making it easier to move away from the set and continue our daily activities. The design of the broadcast time, punctuated by commercials, demands only a part of our attention: we can watch "out of the corner of our eye;" we can pick up a narrative in the middle and comprehend the plot almost immediately; we read, answer the phone, and are often distracted by movements and lights occupying the viewing space. In other words, we do not look at television: we glance at it, pick up the commercial moments, occasionally lose ourselves between private and public selves within our daily life at home.

This experience of television, its homogeneity and lack of regional difference in programming, has established an expectation on the viewer's part of what can and will be available: a standardized packaging of programs under the label of professionalism. The notion of professionalism and the presumed required look of television, the standardized editing, and the breakup of time, with an eye to the viewer's short attention span are wedded to the lowest and most mundane forms of popular culture.[7] The great irony, as new technologies emerge and as television as we know it breaks up into cable, low power, and other alternative systems, is that many independents are following the model of traditional television rather than rethinking its entire support structure and programming content.

The future of television, the extraordinary range of television shapes and sizes from flat screens to miniature sets, the home videorecorder and player, the distribution of discs and cassettes, and off-air subscription programs, all proclaim the potential for the reopening of the telecast and video medium to new viewer networks. This should not simply imply a narrow casting which sees programming to specific viewer interests as representing more sophisticated access to new markets. It does hold the potential to realize what the best video artist has always proclaimed, that we see the medium as a creative and aesthetic medium capable of telecasting a variety of texts which would refashion our visual and cognitive views of image making and narrative meaning. This means that the home viewing space will engage single channel videotapes/discs as continually playing programs viewed as visual art works and as sculptural installation pieces engaging the home space with three-dimensional environments. The example of such video artists as Nam June Paik and Bill Viola, both of whose art has embraced the screen, the television set, and time and image transformation, will help to revise our view of television, the viewing space, and assist the viewer to come to perceive the television screen and program as an open text of possibilities. Television will then be one component in our view of televison-video and not as a simple conduit of information fashioned around a single ideological and phenomenological perspective but as a multi-part discourse of enormous variety and challenge for the viewer.

However utopian the above may sound, it should articulate our goals and make us vigilant to the proposed changes in public access laws for cable and to the reorganization of cable as a new marketing tool. In other words, television is right now being repackaged along new lines supported by the same institutional practices on a federal and corporate level that guided television into the standardized content and formats of the past thirty years. The home viewer "watching television" should not have only single, narrow, options open to him/her. The viewer's choices should be as varied as the producers and senders delivering programs. Then and only then will video and television, the independent producer and the telecast channels, become one in a dialectic of new possibilities and new experiences for the home viewer/producer.

Notes

[1] Ernest Mandel. *Late Capitalism*. London: Verso Edition, 1978.

[2] Erik Barnouw. *Tube of Plenty, The Evolution of American Television*. New York: Oxford University Press, 1975.

[3] Bertolt Brecht "The Radio as an Apparatus of Communication". In *Brecht on Theatre*. New York: Hill and Wang, 1964.

[4] Louis Althussar. "Ideology and Ideological Status Apparatuses (Notes Towards an Investigation)". In *Lenin and Philosophy*. New York: Monthly Review, 1971.

[5] Hans Magnus Enzenberger. *The Consciousness Industry*. New York: Continuum Books, 1974.

[6] Film has been, like photography, available for a much longer time. The public has had hands on contact with both media. Thus there is a familiarity with how film is produced and exhibited. Until only recently, with the introduction of the home video recorder/player, this was not the case with video/television.

[7] The Public Broadcasting System has provided outstanding alternative programming to the traditional fare of commercial television. However, it is for the most part limited to imports and classical culture. Today with the enormous changes in television on the horizon the Public Broadcasting System has apparently given up taking a serious and innovative leadership role.

What makes the world beyond direct experience look natural is a media frame. Certainly we cannot take for granted that the world depicted is simply the world that exists. Many things exist. At each moment the world is rife with events. Even within a given event there is an infinity of noticeable details. Frames are principles of selection, emphasis, and presentation composed of little tacit theories about what exists, what happens, and what matters.

THE WHOLE WORLD IS WATCHING

Todd Gitlin

SINCE THE ADVENT OF RADIO BROADCASTING half a century ago, social movements have organized, campaigned, and formed their social identities on a floodlit social terrain. The economic concentration of the media and their speed and efficiency in spreading news and telling stories have combined to produce a new situation for movements seeking to change the order of society. Yet movements, media, and sociology alike have been slow to explore the meanings of modern cultural surroundings.

People directly know only tiny regions of social life; their beliefs and loyalties lack deep tradition. The modern situation is precisely the common vulnerability to rumor, news, trend, and fashion: lacking the assurances of tradition, or of shared political power, people are pressed to rely on mass media for bearings in an obscure and shifting world. And the process is reciprocal: pervasive mass media help pulverize political community, thereby deepening popular dependence on the media themselves. The media bring a manufactured public world into private space. From within their private crevices, people find themselves relying on the media for concepts, for images of their heroes, for guiding information, for emotional charges, for a recognition of public values, for symbols in general, even for language. Of all the institutions of daily life, the media specialize in orchestrating everyday consciousness—by virtue of their pervasiveness, their accessibility, their centralized symbolic capacity. They name the world's parts, they certify reality *as* reality—and when their certifications are doubted and opposed, as they surely are, it is those same certifications that limit the terms of effective opposition. To put it simply: the mass media have become core systems for the distribution of ideology.

That is to say, every day, directly or indirectly, by statement and omission, in pictures and words, in entertainment and news and advertisement, the mass media produce fields of definition and association, symbol and rhetoric, through which ideology becomes manifest and concrete. One important task for ideology is to define—and also define away—its opposition. This has always been true, of course. But the omnipresence and centralization of the mass media, and their integration into the dominant economic sector and the web of the State, create new conditions for opposition. The New Left of the 1960s, facing nightly television news, wire service reports, and a journalistic ideology of "objectivity," inhabited a cultural world vastly different from that of the Populist small farmers' movement of the 1890s, with its fifteen hundred autonomous weekly newspapers, or that of the worker-based Socialist Party of the early 1900s, with its own newspapers circulating in the millions. By the sixties, American society was dominated by a *consolidated* corporate economy, no longer by a nascent one. The dream of Manifest Destiny had become realized in a missile-brandishing national security state. And astonishingly, America was now the first society in the history of the world with more college students than farmers. The social base of radical opposition, accordingly, had shifted—from small farmers and immigrant workers to blacks, students, youth, and women. What was transformed was not only the dominant *structures* of capitalist society, but its *textures*. The whole quality of political movements, their procedures and tones, their cultural commitments, had changed. There was now a mass market culture industry, and opposition movements had to reckon with it—had to operate on its edges, in its interstices, and against it. The New Left, like its Populist and Socialist Party predecessors, had its own scatter of "underground" newspapers, with hundreds of thousands of readers, but every night some twenty million Americans watched Walter Cronkite's news, an almost equal number watched Chet Huntley's and David Brinkley's, and over sixty million bought daily newspapers which purchased most of their news from one of two international wire services. In a floodlit society, it becomes extemely difficult, perhaps unimaginable, for an opposition movement to define itself and its world view, to build up an infrastructure of self-generated cultural institutions, outside the dominant culture.[1] Truly, the process of making meanings in the world of centralized commercial culture has become comparable to the process of making value in the world through labor. Just as people *as workers* have no voice in what they make, how they make it, or how the product is distributed and used, so do people *as producers of meaning* have no voice in what the media make of what they say or do, or in the context within which the media frame their activity. The resulting meanings, now mediated, acquire an eery substance in the real world, standing outside their ostensible makers and confronting them as an alien force. The social meanings of intentional action have been deformed beyond recognition.

In the late twentieth century, political movements feel called upon to rely on large-scale communications in order to *matter*, to say who they are and what they intend to publics they want to sway; but in the process they become "newsworthy" only by submitting to the implicit rules of newsmaking, by conforming to journalistic notions (themselves embedded in history) of what a "story" is, what an "event" is, what a "protest" is. The processed image then tends to *become* "the movement" for wider publics and institutions who have few alternative sources of information, or none at all, about it; that image has its impact on public policy, and when the movement is being opposed, what is being opposed is in large part a set of mass-mediated images. Mass media define the public significance of movement events or, by blanking them out, actively deprive them of larger significance. Media images also become implicated in a movement's self-image; media certify leaders and officially noteworthy "personalities;" indeed, they are able to convert leadership into *celebrity*, something quite different. The forms of coverage accrete into systematic framing, and this framing, much amplified, helps determine the movement's fate.

For what defines a movement as "good copy" is often flamboyance, often the presence of a media-certified celebrity-leader, and usually a certain fit with whatever frame the newsmakers have construed to be "the story" at a given time; but these qualities of the image are not what movements intend to be their projects, their identities, their goals. Yet while they constrict and deform movements, the media do amplify the issues which fuel these same movements; they expose scandal in the State and in the corporations, while reserving to duly constituted authority the legitimate right to remedy evils. The liberal media quietly invoke the need for reform—while disparaging movements radically opposed to the system that needs reforming.

The routines of journalism, set within the economic and political interests of news organizations, normally and regularly combine to select certain versions of reality over others. Day by day, *normal* organizational procedures define "the story," identify the protagonists and the issues, and suggest appropriate attitudes toward them. Only episodically, in moments of political crisis and large-scale shifts in the overarching hegemonic ideology, do political and economic managers and owners intervene directly to regear or reinforce the prevailing journalistic routines. But most of the time the take-for-granted code of "objectivity" and "balance" presses reporters to seek out scruffy-looking, chanting, "Viet Cong" flag-waving demonstrators and to counterpose them to reasonable-sounding, fact-brandishing authorities. Calm and cautionary tones of voice affirm that all "disturbance" is or should be under control by rational authority; code words like *disturbance* commend the established order. Hotheads carry on, the message connotes, while wiser heads, officials and reporters both, with superb self-control, watch the unenlightened ones make trouble.

Yet these conventions originate, persist, and shift in historical time. The world of news production is not self-enclosed; for commercial as well as professional reasons, it cannot afford to ignore big ideological changes. Yesterday's ignored or ridiculed kook becomes today's respected "consumer activist," while at the same time the mediated image of the wild sixties yields to the image of laid-back, apathetic, self-satisfied seventies. Yesterday's revolutionary John Froines of the Chicago Seven, who went to Washington in 1971 to shut down the government, goes to work for it in 1977 at a high salary; in 1977, Mark Rudd surfaces from the Weather Underground, and the sturdy meta-father Walter Cronkite chuckles approvingly as he reports that Mark's father thinks the age of thirty is "too old to be a revolutionary"—these are widely publicized signs of presumable calmer, saner times. Meanwhile, movements for utility rate reform, for unionization in the South, for full employment, for disarmament, and against nuclear power—movements which are not led by "recognized leaders" (those whom the media selectively acknowledged as celebrities in the first place) and which fall outside the prevailing frames ("the New Left is dead," "America is moving to the right")—are routinely neglected or denigrated—until the prevailing frame changes (as it did after the accident at Three Mile Island). An activist against nuclear weapons, released from jail in May 1978 after a series of demonstrations at the Rocky Flats, Colorado, factory that manufactures plutonium triggers for all American H-bombs, telephoned an editor he knew in the *New York Times's* Washington bureau to ask whether the *Times* had been covering these demonstrations and arrests. No, the editor said, adding: "America is tired of protest. America is tired of Daniel Ellsberg." Blackouts do take place; the editorial or executive censor rationalized his expurgation, condescendingly and disingenuously, as the good shepherd's fairminded act of professional news judgement, as his service to the benighted, homogenized, presumable sovereign audience. The closer an issue is to the core interests of national political elites, the more likely is a blackout of news that effectively challenges that interest. That there is safety in the country's nuclear weapons program is, to date, a core principle; and so news of its menace is extremely difficult to get reported—far more difficult, for example, than news about dangers of nuclear power after Three Mile Island. But if the issue is contested at an elite level, or if an elite position has not yet crystallized, journalism's more regular approach is to *process* social opposition, to control its image and to diffuse it at the same time, to absorb what can be absorbed into the dominant structure of definitions and images and to push the rest to the margins of social life.

The processed message becomes complex. To take a single example of a news item: on the CBS Evening News of May 8, 1976, Dan Rather reported that the FBI's burglaries and wiretaps began in the thirties and continued through World War II and the Cold War; and he concluded the piece by saying that these activities reached a peak "during the civil disturbances of the sixties." In this piece we can see some of the contradictory workings of broadcast journalism—and the limits within which contradictory forces play themselves out. First of all, Rather was conveying the information that a once sacrosanct sector of the State had been violating the law for decades. Second, and more subtly—with a clipped, no-nonsense manner and a tough-but-gentle, trustworthy, Watergate-certified voice of technocracy—he was deploring this law-breaking, lending support to those institutions within the State that brought it to the surface and now proposed to stop it, and affirming that the media are integral to this self-correcting system as a whole. Third, he was defining a onetime political opposition *outside* the State as "civil disturbance." The black and student opposition movements of the sixties, which would look different if they were called, say, "movements for peace and justice," were reduced to nasty little things. Through his language, Rather was inviting the audience to identify with forces of reason within the state: with the very source of the story, most likely. In a single news item, with (I imagine) no deliberate forethought, Rather was (a) identifying an abuse of government, (b) legitimating reform within the existing institutions, and (c) rendering illegitimate popular or radical opposition outside the State. The news that man has bitten dog carries an unspoken morality: it proposes to coax men to stop biting those particular dogs, so that the world can be restored to its essential soundness. In such quiet fashion, not deliberately, and without calling attention to this spotlighting process, the media divide movements into legitimate main acts and illegitimate sideshows, so that these disctinctions appear "natural," matters of "common sense."[2]

What makes the world beyond direct experience look natural is a media *frame*.[3] Certainly we cannot take for granted that the world depicted is simply the world that exists. Many things exist. At each moment the world is rife with events. Even within a given event there is an infinity of noticeable details. Frames are principles of selection, emphasis, and presentation composed of little tacit theories about what exists, what happens, and what matters. In everyday life, as Erving Goffman has amply demonstrated, we frame reality in order to negotiate it, manage it, comprehend it, and choose appropriate repertories of cognition and action.[4] *Media* frames, largely unspoken and unacknowledged, organize the world both for journalists who report it and, in some important degree, for us who rely on their reports. *Media frames are persistent patterns of cognition, interpretation, and presentation, of selection, emphasis, and exclusion, by which symbol-handlers routinely organize discourse, whether verbal or visual.* Frames enable journalists to process large amounts of information quickly and routinely: to recognize it as information, to assign it to cognitive categories, and to package it for efficient relay to their audiences. Thus, for organizational reasons alone, frames are unavoidable, and journalism is organized to regulate their production. Any analytic approach to journalism—indeed, to the production of any mass-mediated content—must ask: What is the frame here? Why this frame and not another? What patterns are shared by the frames clamped over this event and the frames clamped over that one, by frames in different media in different places at different moments? And how does the news-reporting institution regulate these regularities?

And then: What difference do the frames make for the larger world?

The issue of the influence of mass media on larger political currents does not, of course, emerge only with the rise of broadcasting. In the Paris of a century and a half ago, when the commercial press was young, a journalistic novice and littérateur-around-town named Honoré de Balzac was already fascinated by the force of commercialized images. Central to his vivid semi-autobiographical novel, *Lost Illusions*, was the giddy, corroded career of the journalist. Balzac saw that the press degraded writers into purveyors of commodities. Writing in 1839 about the wild and miserable spectacle of "A Provincial Great Man in Paris," Balzac in one snatch of dinner-party dialogue picked up the dispute aborning over political consequences of a mass press; he was alert to the fears of reactionaries and the hopes of Enlightenment liberals alike:

"The power and influence of the press are only at their dawn," said Finot. "Journalism is in its infancy, it will grow and grow. Ten years hence everything will be subjected to publicity. Thought will enlighten everything, it—"

"Will blight everything," interposed Blondet.

"That's a *bon mot,*" said Claude Vignon.

"It will make kings," said Lousteau.

"It will unmake monarchies," said the diplomat.

"If the press did not exist," said Blondet, "we could get along without it; but it's here, so we live on it."

"You will die of it," said the diplomat. "Don't you see that the superiority of the masses, assuming that you enlighten them, would make individual greatness the more difficult of attainment; that, if you sow reasoning power in the heart of the lower classes, you will reap revolution, and that you will be the first victims?"[5]

Balzac's ear for hopes and fears and new social tensions was acute; he was present at the making of a new institution in a new social era. Since then, of course, radio and now television have become standard home furnishings. And in considerable measure broadcast content has become part of the popular ideological furniture as well. But while researchers debate the exact "effects" of mass media on the popularity of presidential candidates and presidents, or the "effects" on specific patterns of voting or the salience of issues, evidence quietly accumulates that the texture of political life has changed since broadcasting became a central feature of American life. Media certainly help set the agendas for political discourse; although they are far from autonomous, they do not passively reflect the agendas of the State, the parties, the corporations, or "public opinion."[6] The centralization and commercialization of the mass media of communication make them instruments of cultural dominance on a scale unimaged even by Balzac. In some ways the very ubiquity of the mass media removes media *as a whole system* from the scope of positivist social analysis; for how may we "measure" the "impact" of a social force which is omnipresent within social life and which has a great deal to do with constituting it? I work from the assumption that the mass media are, to say the least, a significant social force in the forming and delimiting of public assumptions, attitudes, and moods—of ideology, in short. They sometimes generate, sometimes amplify a field of legitimate discourse that shapes the public's "definitions of its situations," and they work through selections and omissions, through emphases and tones, through all their forms of treatment.

Such ideological force is central to the continuation of the established order. I take it for now that the central command structures of this order are an oligopolized, privately controlled corporate economy and its intimate ally, the bureaucratic national security state, together embedded within a capitalist world complex of nation-states. But the economic and political powers of twentieth-century capitalist society, while formidable, do not by themselves account for the society's persistence, do not secure the dominant institutions against the radical consequences of the system's deep and enduring conflicts. In the language of present-day social theory, why does the population accord legitimacy to the prevailing institutions? The goods are delivered, true; but why do citizens agree to identify themselves and to behave as consumers, devoting themselves to labor in a deteriorating environment in order to acquire private possessions and services as emblems of satisfaction? The answers are by no means self-evident. But however we approach these questions, the answers will have to be found in the realm of ideology, of culture in the broadest sense. Society is not a machine or a thing; it is a coexistence of human beings who do what they do (including maintaining or changing a social structure) as sentient, reasoning, moral, and active beings who experience the world, who are not simply "caused" by it. The patterned experiencing of the world takes place in the realm of what we all ideology. And any social theory of ideology asks two interlocking questions: How and where are ideas generated in society? And why are certain ideas accepted or rejected in varying degrees at different times?

In the version of Marxist theory inaugurated by Antonio Gramsci, *hege-*

mony is the name given to a ruling class's domination through ideology, through the shaping of popular consent.[7] More recently, Raymond Williams has transcended the classical Marxist base-superstructure dichotomy (in which the "material base" of "forces and relations of production" "gives rise" to the ideological "superstructure"). Williams has proposed a notion of hegemony as "not only the articulate upper level of 'ideology,'" but "a whole body of practices and expectations" which "constitutes a sense of reality for most people in the society."[8] The main economic structures, or "relations of production," set limits on the ideologies and commonsense understandings that circulate as ways of making sense of the world—without mechanically "determining" them. The fact that the networks are capitalist corporations, for example, does not automatically decree the precise frame of a report on socialism, but it does preclude continuing, emphatic reports that would embrace socialism as the most reasonable framework for the solution of social problems. One need not accept all of Gramsci's analytic baggage to see the penetrating importance of the notion of hegemony—uniting persuasion from above with consent from below—for comprehending the endurance of advanced capitalist society. In particular, one need not accept a strictly Marxist premise that the "material base" of "forces of production" in *any* sense (even "ultimately") precedes cultures.[9] But I retain Gramsci's core conception: those who rule the dominant institutions secure their power in large measure directly *and indirectly*, by impressing their definitions of the situation upon those they rule and, if not usurping the whole of ideological space, still significantly limiting what is thought throughout the society. The notion of hegemony that I am working with is an active one: hegemony operating through a complex web of social activities and institutional procedures. Hegemony is done by the dominant and collaborated in by the dominated.

Hegemonic ideology enters into everything people do and think is "natural"—making a living, loving, playing, believing, knowing, even rebelling. In every sphere of social activity, it meshes with the "common sense" through which people make the world seem intelligible; it tries to *become* that common sense. Yet, at the same time, people only partially and unevenly accept the hegemonic terms; they stretch, dispute, and sometimes struggle to transform the hegemonic ideology. Indeed, its contents shift to a certain degree, as the desires and strategies of the top institutions shift, and as different coalitions form among the dominant social groups; in turn, these desires and strategies are modified, moderated by popular currents. In corporate capitalist society (and in state socialism as well), the schools and the mass media specialize in formulating and conveying national ideology. At the same time, indirectly, the media—at least in liberal capitalist society—take account of certain popular currents and pressures, symbolically incorporating them, repackaging and distributing them throughout the society. That is to say, groups out of power—radical students, farm workers, feminists, environmentalists, or homeowners groaning under the property tax—can contest the prevailing structures of power and definitions of reality. One strategy which insurgent social movements adopt is to make "news events."

The media create and relay images of order. Yet the social reality is enormously complex, fluid, and self-contradictory, even in its own terms. Movements constantly boil up out of the everyday suffering and grievance of dominated groups. From their sense of injury and their desire for justice, movements assert their interests, mobilize their resources, make their demands for reform, and try to find space to live their alternative "lifestyles." These *alternative* visions are not yet *oppositional*—not until they challenge the main structures and ideas of the existing order: the preeminence of the corporate economy, the militarized State, and authoritarian social relations as a whole. In liberal capitalist society, movements embody and exploit the fact that the dominant ideology enfolds contradictory values: liberty versus equality, democracy versus hierarchy, public rights versus property rights, rational claims to truth versus the arrogations and mystifications of power.[10] Then how does enduring ideology find its way into the news, absorbing and ironing out contradictions with relative consistency? How, in particular, are rather standardized frames clamped onto the reporting of insurgent movements? For the most part, through journalists' routines.

These routines are structured in the ways journalists are socialized from childhood, and then trained, recruited, assigned, edited, rewarded, and promoted on the job; they decisively shape ways in which news is defined, events are considered newsworthy, and "objectivity" is secured. News is managed routinely, automatically, as reporters import definitions of newsworthiness from editors and institutional beats, as they accept the analytical frameworks of officials even while taking up adversary positions. When reporters make decisions about what to cover and how, rarely do they deliberate about ideological assumptions or political consequences.[11] Simply by doing their jobs, journalists tend to serve the political and economic elite definitions of reality.

But there are disruptive moments, critical times when the routines no longer serve a coherent hegemonic interest. The routines produce news that no longer harmonizes with the hegemonic ideology, or with important elite interests as the elites construe them; or the elites are themselves so divided as to quarrel over the content of the news. (In the extreme case, as in Chile in 1973, the hegemonic ideology is pushed to the extremity of its self-contradiction, and snaps; the dominant frame then shifts dramatically, in that case toward the Right.) At these critical moments, political and economic elites (including owners and executives of media corporations) are more likely to intervene directly in journalistic routine, attempting to keep journalism within harness. To put it another way, the cultural apparatus normally maintains its own momentum, its own standards and procedures, which grant it a certain independence from top political and economic elites. In a liberal capitalistic society, this bounded but real independence helps legitimate the institutional order as a whole and the news in particular. But the elites prefer not to let such independence stretch "too far." It serves the interests of the elites as long as it is "relative," as long as it does not violate core hegemonic values or contribute too heavily to radical critique or social unrest. (It is the elites who determine, or establish routines to determine, what goes "too far.") Yet when elites are themselves at odds in important ways, and when core values are deeply disputed—as happened in the sixties—journalism itself becomes contested. Opposition groups pressing for social and political change can exploit self-contradictions in hegemonic ideology, including its journalistic codes. Society-wide conflict is then carried into the cultural institutions, though in muted and sanitized forms. And then ideological domestication plays an important part—along with the less visible activities of the police[12]—in taming and isolating ideological threats to the system. . . .

News is one component of popular culture; the study of news should ultimately be enfolded within a more ample study of all the forms of cultural production and their ideology. Television entertainment is also an ideological field, and must have played a part in formulating and crystallizing the cultural tendencies of the sixties; surely it deserves extensive treatment of its own.[13] So do other cultural forms, including popular songs, popular fiction (genre novels as well as magazine stories), jokes, and popular films (which are not necessarily the acclaimed films which critics prefer to see and to analyze); so do the careers of pop stars like Bob Dylan and Joan Baez, the San Francisco bands and hip heroes who stood somewhere on the thin and fluid boundary between the New Left and the counterculture. Let popular culture have its analytic due: we live in it.

Notes

[1] This point is made by Walter Adamson, "Beyond Reform and Revolution: Notes on Political Education in Gramsci, Habermas and Arendt," *Theory and Society* 6(November 1978): 429-460.

[2] For further analysis of the meaning of this and other television news items, see Todd Gitlin, "Spotlights and Shadows: Television and the Culture of Politics," *College English* 38(April 1977): 791-796.

[3] On media frames, see Gaye Tuchman, *Making News* (New York: The Free Press, 1978); and Stuart Hall, "Encoding and Decoding in the Television Discourse," mimeographed paper, Centre for Contemporary Cultural Studies, University of Birmingham, England, 1973.

[4] Erving Goffman, *Frame Analysis: An Essay on the Organization of Experience* (New York: Harper and Row, 1974), pp. 10-11 and passim.

[5] Honoré de Balzac, *Lost Illusions*, trans. G. Burnham Ives (Philadelphia: George Barrie, 1898), Vol. 2, p.112.

[6] Least of all, public opinion. Evidence is accumulating that the priorities conveyed by the media in their treatment of political issues lead public opinion rather than following it. See Maxwell E. McCombs and Donald I. Shaw, "The Agenda-Setting Function of Mass Media," *Public Opinion Quarterly* 36 (1972): 176-187; Jack M. McLeod, Lee B. Becker, and James E. Byrnes, "Another Look at the Agenda-Setting Function of the Press," *Communication Research* 1(April 1974): 131-166; Lee B. Becker, Maxwell E. McCombs, and Jack M. McLeod, "The Development of Political Cognitions," in Steven H. Chaffee, ed., *Political Communication* (Beverly Hills, Calif.: Sage Publications, 1975), pp. 21-63, especially pp. 38-53; and Jay G. Blumler and Denis McQuail, *Television in Politics: Its Uses and Influence (London: Faber, 1968).*

[7] *Antonio Gramsci, Selections from the Prison Notebooks*, ed. and trans. Quintin Hoare and Geoffrey Nowell Smith (New York: International, 1971).

[8] Raymond Williams, "Base and Superstructure in Marxist Cultural Theory," *New Left Review*, No. 82 (1973), pp. 3-16. See also Williams, *Marxism and Literature* (New York: Oxford University Press, 1977), especially pp. 108-114.

[9] For a brilliant demonstration of ways in which culture helps *constitute* a given society's "material base," and in particular the way in which the bourgeois concept of utility conditions capitalism's claims to efficiency, see Marshall Sahlins, *Culture and Practical Reason* (Chicago: University of Chicago Press, 1976), Part 2.

[10] I adapt this argument from my "Prime Time Ideology: The Hegemonic Process in Television Entertainment," *Social Problems* 26 (February 1979): 264-265.

[11] As Gaye Tuchman writes, "news both draws upon and reproduces institutional structures" *(Making News*, p. 210). For particulars, see Leon V. Sigal, *Reporters and Officials: The Organization and Politics of Newsmaking* (Lexington, Mass.: D.C. Heath, 1973);

Bernard Roshco, *Newsmaking* (Chicago: University of Chicago Press, 1975); and most fully, Herbert Gans, *Deciding What's News* (New York: Pantheon, 1979).

[12]Very little has been written on direct relations between police agencies and mass media. Gans *(Deciding What's News,* p. 121) makes the valuable point that "perhaps the most able sources are organizations that carry out the equivalent of investigative reporting, offer the results of their work as 'exclusives,' and can afford to do so anonymously, foregoing the rewards of publicity." For a survey of the FBI's COINTELPRO media operations, especially in New York, Chicago, Los Angeles, and Milwaukee, at least between 1956 and 1971, and a few extant details of direct cooperation between the FBI and reporters, see Chip Berlet, "COINTELPRO: What the (Deleted) Was It? Media Op," *The Public Eye 1* (April 1978): 28-38. I know of no evidence of cooperation between the FBI and either CBS News or the *New York Times,* but this entire field is *terra incognita.*

[13] Considering the great amount of time Americans and others spend watching TV entertainment, there is a great imbalance in sociological attention: many more studies have been done of the production and meanings of news, which is more transparently available for political understandings, than of everyday fiction. I have sketched some preliminary categories for the analysis of TV entertainment conventions in "Prime Time Ideology," (note 10, above). On TV entertainment and its evolution in general, there is abundant material in Erik Barnouw's *The Image Empiref (New York: Oxford University Press, 1970), Tube of Plenty* (New York: Oxford University Press, 1975), and *The Sponsor* (New York: Oxford University Press, 1978); in Fiske and Hartley, *Reading Television*, and Hall, "Encoding and Decoding in the Television Discourse," both cited in note 13, above; and in Rose K. Goldsen, *The Show and Tell Machine* (New York: Dial Press, 1977), though many analytic questions remain. On the content and history of specific shows and types of shows, see Danny Czitrom, "Bilko: A Sitcom for All Seasons," *Cultural Correspondence* 4 (Spring 1977): 16-19; Todd Gitlin, "The Televised Professional," *Social Policy*, November-December 1977, pp. 94-99; Pete Knutson, "Dragnet: The Perfect Crime?" *Liberation* 18 (May 1974): 28-31; Michael R. Real, *Mass-Mediated Culture* (Englewood Cliffs N.J.: Prentice-Hall, 1977), pp. 118-139 (on medical shows); and Bob Schneider, "Spelling's Salvation Armies," *Cultural Correspondence*, No. 4 (Spring 1977): 27-36 (on police shows). On soap operas, see Dennis Porter, "Soap Time: Thoughts on a Commodity Art Form," *College English* 38 (April 1977): 782-788. On the production process, see Muriel G. Cantor, *The Hollywood TV Producer* (New York: Basic Books, 1971); Les Brown, *Television: The Business Behind the Box* (New York: Harcourt Brace Jovanovich, 1971); and Gaye Tuchman, "Assembling a Network Talk-Show," in Tuchman, ed., *The TV Establishment* (Englewood Cliffs, N.J.: Prentice-Hall, 1974), pp. 119-135.

MARGARET MEAD AND THE SHIFT FROM "VISUAL ANTHROPOLOGY" TO THE "ANTHROPOLOGY OF VISUAL COMMUNICATION"

Sol Worth

I WOULD LIKE, in this discussion, to explore a shift in how certain problems in the study of culture have come to be conceptualized.[1] These problems may best be understood by examining how one label, "visual anthropology," led to the creation of another, "the anthropology of visual communication." In order to delineate and examine some of the arguments, problems, and methods involved in this shift, it will be helpful for me to cite, and to use as my explanatory fulcrum, the work as well as the persona of Margaret Mead.

I am doing this on an occasion meant to honor her, but am aware that even that act, as so often happens with Dr. Mead, inevitably gets mixed up with a review of the history and problems in communications and anthropology. I should add that I am aware that even as we try to develop a history in this field, we also *are* in many ways that same history.

To introduce some of these issues in the history of communications study, let me quote from an informant whose comments and life history may lay the groundwork for certain of the problems I will be talking about. Some of you may still remember a television series of several years ago called "An American Family." It consisted of twelve one-hour film presentations. One of the major participants in that visual event was Mrs. Patricia Loud, the mother of that "American" family. In a letter to some of her acquaintances which she subsequently made public, Mrs. Loud wrote:

Margaret Mead, bless her friendly voice, has written glowingly that the series constituted some sort of breakthrough, a demonstration of a new tool for use in sociology and anthropology. Having been the object of that tool, I think I am competent to say that it won't work.

Later in her letter she continues:

Like Kafka's prisoner, I am frightened, confused. . . . I find myself shrink-

ing in defense, not only from critics and detractors, but from friends, sympathizers and, finally, myself. . . . The truth is starting to dawn on me that we have been ground through the big media machine and are coming out entertainment. The treatment of us as objects and things instead of people has caused us wildly anxious days and nights. But I would do it again if, in fact, I could just be sure that it did what the producer said it was supposed to do. If we failed, was it because of my family, the editing, the publicity, or because public television doesn't educate? If we failed, what role did the limitations of film and TV tape play? Can electronic media really arouse awareness and critical faculties? Did we, family and network alike, serve up great slices of ourselves—irretrievable slices—that only serve to entertain briefly, to titillate, and diminish into nothing? —*(Loud, 1974)*

Margaret Mead did not photograph, edit, or produce this visual event that Pat Loud speaks of. But in ways that I will describe, she can be understood to be a major influence in this and other attempts to show a family in the context of television. More importantly, her work over the past fifty years can help us to understand many of the questions that Pat Loud's cry of distress raised for her.

There are, it seems to me, at least three basic premises which Mrs. Loud's letter forces us to examine. First is our deeply held and largely unexamined notion that all or most photographs, and, in particular, motion pictures, are a mirror of the people, objects, and events that these media record photochemically. Second is the questionable logic of the jump we make when we say that the resultant photographic image could be, should be, and most often is something called "real," "reality," or "truth." A third concern, which is central to Pat Loud personally, and increasingly to all people studied or observed by cameras for television, whether for science, politics, or art, is the effect of being, as she puts it, "the object of that tool."

When "An American Family" was first shown on American television in 1972, mass media critics, psychoanalysts, sociologists, and historians, as well as *Time, Newsweek,* and the *New York Times,* felt compelled to comment. Almost all—except Margaret—expressed dismay, upset, and even anger over the series. Many of these strong feelings were no doubt occasioned by the films themselves—by the way they were advertised and presented, as well as by the events depicted in them. But much of the upset was also caused, I believe, by the fact that Margaret Mead said publicly, and with approval, that this notion of depicting a family on television was a worthwhile, revolutionary, daring, and possibly fruitful step in the use of the mass media. She even compared the idea of presenting a family on television to the idea of the novel, suggesting that it might, if we learned to use it, have a similar impact upon the culture within which we live.

Interestingly enough, in October 1976 the United Church of Christ, the Public Broadcasting System and Westinghouse Television will present a series titled "Six Families," in which the same thing that was tried in the Loud family series will now be done on comparative basis. It seems that most of the objections of social scientists to the Loud family series were that this use of "real" people on TV was unethical, immoral, and indecent. It made, many people argued, a nation of prurient Peeping Toms out of the American people. It is, of course, "the church" which in our society can take initiative and argue that an examination of how people live, shown on TV, is not only not Peeping Tomism but the most moral kind of act for a mass medium. We will have to see whether social scientists, TV critics, and newspapers will even notice this second instance of the presentation of an American family on television.

The problem for those who heard or read what Margaret Mead said about this new use of film, whether they were academics, newspaper people, or even subjects, was that we were just beginning to understand what Bateson and Mead had said in 1942. We were just beginning to accept the idea that photographs could be taken and used seriously, as an artistic as well as a scientific event. We were not ready to acknowledge that we were beyond the point of being excited by the fact that a camera worked at all. It was, after all, understood as early as 1900 that photographs and motion pictures could be more than a record of data and that they were always less than what we saw with our eyes. Let us look at how it started.

The first set of photographs called motion pictures was made by Eadweard Muybridge in 1877, as scientific evidence of a very serious kind. He invented a process for showing things in motion in order to settle a bet for Governor Leland Stanford of California about whether horses had all four feet off the ground when they ran at a gallop. Our popular myth about cinema and truth started here. If the motion picture camera showed it—everyone seemed to, and wanted to, believe—it had to be so. Edison in the United States and Lumière in Europe invented more sophisticated machines for taking motion pictures, and, interestingly enough, the first films made with those primitive motion picture cameras between 1895 and 1900 had much of the spirit of what is still called ethnographic filming. They presented what the early filmmakers advertised as "the world as it really was." Lumiere's first film showed French workers in the Peugeot auto factory outside Paris lining up to punch a time clock. Edison's first film showed his assistant in the act of sneezing. Both Edison and Lumière went on from there to depict other "real" and "documentary" scenes of people walking in the street, bathing at the beach, eating, embarking on a train, and so on.

The issue of reality in film was already being argued in 1901, not by scientists or artists but by film manufacturers. The Riley Brothers catalogue of 1900-1901 states:

The films listed here are the very best quality. They are clean and sharp and full of vigor. They are properly treated in the course of manufacture and do not leave the celluloid. None of the subjects have been "faked." All are genuine photographs taken without pre-arrangement *and are consequently most natural.*

The notion of a systematically made ethnographic record of the geographic and physical environment of a city, in a style conforming to ideas promulgated by Collier (1967), was also being advertised and sold in 1901. The Edison catalogue for that year states:

New York in a Blizzard. *Our camera is revolved from right to left and takes in Madison Square, Madison Square Garden, looks up Broadway from south to north, passes the Fifth Avenue Hotel and ends looking down Twenty-third Street.*

Such a film could have been made with an ethnographic soundtrack on instructions given to modern ethnofilmmakers by archivists in the United States and several countries in Western Europe.

We have, it seems, come a long way from the days when just being able to make a picture (moving or still) of strange or familiar people in our own country or people in far-away lands doing exotic things was excuse enough for lugging a camera to the field or to our living rooms. In those earlier times, from 1895 to about 1920, the term *visual anthropology* had not yet been coined. People just took pictures, most often to "prove" that the people and places they were lecturing about or studying actually existed. In some cases, they took pictures so that when they returned to their own homes they could study, in greater detail and with more time, what these people and things looked like. Archaeologists quite early (around 1900) began to use this new miracle machine. They found the camera not only quicker than making copy drawings of the artifacts they uncovered, but more accurate—truer to life, or the artifact. I believe that it was from the use to which archaeologists put photographs that cultural anthropology developed its first, and still extremely important, conceptual paradigm about the use of pictures: that the purpose of taking pictures in the field is to show the "truth" about whatever it was the picture purported to be of: an arrowhead, a potsherd, a house, a person, a dance, a ceremony, or any other behavior that people could perform, and cameras record, in the same spatial frame. The subtle shift that took place when we expanded on the role of photography in anthropology and archaeology, from the use of a photograph of an arrowhead or a potsherd as evidence of existence to the use of a photograph of people as evidence of human behavior, is a particularly important and unexamined aspect of the history of social science, and especially of the history of anthropology.

A conceptual difficulty that we now face results from the fact that the avowals of truth in photography made in the 1901 film catalogues now seem self-evident to us. In fact, a major problem in thinking about the use of photography in social science today is not that photographs are not true, but that that is not the purpose we use them for. One of the clearest expressions of this dilemma, and one that shaped much of my own thinking about the uses of photography in social science, can be found in the appendix to *Growth and Culture* (1951), by Mead and MacGregor, based on the photographic work of Bateson and Mead in Bali. Mead writes:

Anthropological field work is based upon the assumption that human behavior is systematic. . . that in such research the principal tool is consciousness of pattern (and) that the anthropologist brings to this work a training in the expectation of form.

Mead then explains how the photographs taken in Bali were used. Of some 25,000 still pictures taken by Bateson, 4,000 were chosen, from which MacGregor, Mead, and the Gesell group could find a set of patterns derived from a study of photographs—not from the photographs themselves—which could then be compared with patterns found in the study of American children. It is important to emphasize Mead's subtle but powerful distinction: the patterns of behavior in this case were derived from the study and analysis of the photographs, not from the photographs as a magic mirror of pattern. Mead states quite clearly: "These photographs are designed not to *prove,* but to *illustrate. . . .*"

In effect, what Mead has been trying to teach us is what one of her teachers, Ruth Benedict, taught her: "patterns of culture" are what we are presenting when we do anthropology; and taking photographs, or looking, or taking notes are tools for articulating and stating the patterns that we, as anthropologists, wish to show to others. It is that old lesson about culture which we seem not to understand as it affects our use of the photograph. Somehow, our myth system about photographs helps us to forget that the photograph is not the pattern. Somehow we tend to think of a photograph not as something we use—as evidence, to illustrate pattern, to inform ourselves, or to make statements with—but as something we call "truth" or "reality."

One should distinguish between the photograph as a record *about* culture and the photograph as a record *of* culture. One should also distinguish between using a medium and studying how a medium is used. In terms of the camera, the distinction I want to emphasize is that between the scientists' use of the camera as a tool to collect data about culture, and studying how the camera is used by members of a culture. This distinction is, I feel, central to understanding the work done with this medium of communication in the last eighty years. On one level, the photograph is an aide-memoire to the scientist, equal to his pencil, notebook, or typewriter. It is not—as we now know, from recent work by Chalfen (1975), Ruby (1975), and others—merely a bunch of snapshots or home movies made by an anthropologist. In the hands of well-trained observers, it has become a tool for recording, not the truth of what is *out there*, but the truth of what is *in there*, in the anthropologist's mind, as a trained observer puts observations of "out there" on record. Photography, as a record about culture, spans the distance from the casual snapshot, which reminds one of what a house or an informant looked like, to the systematic work of a Mead, a Bateson, or a Birdwhistell. And here I must emphasize that it is not their photography that is important, but their analysis of it. The reason their photographs and films are records is that they were taken in ways which allowed them to be analyzed so as to illustrate patterns observed by scientists who knew what they were looking for.

Let us now turn to the second level of analysis of photographs and films as records of culture—as objects and events which can be studied in the context of the culture within which they were used. The photographs and films analyzed in this way are understood to be parts of culture in their own right, just as conversations, novels, plays, and other symbolic behavior have been understood to be. Here I am talking about looking at how someone takes a photograph or puts together an advertisement, as well as how he makes a movie. One is concerned at this level, for example, with finding patterns of moviemaking used by anthropologists, physicists, and Hollywood entrepreneurs, by college students, by "artists," by people using 8-mm. cameras in our own culture, as well as by Navajo Indians or members of any other group who are making photographs or movies for purposes of their own.

Here one looks for patterns dealing with, for example, what can be photographed and what cannot, what content can be displayed, was actually displayed, and how that display was organized and structured. Was it arranged according to how these people tell stories? To how they speak, or to the very language and grammar that they use? Recent work by one of my students, Earl Higgins, seems to indicate that even among the congenitally deaf, the "grammar" and related patterns of their sign language influence how speakers of American Sign Language structure films that they make.

Here again, although Margaret Mead was not the first to think of examining photography and films in this way, she articulated the ideas and related them to an understanding of culture in a larger and systematic way. Mead, in *The Study of Culture at a Distance* (Mead and Metraux 1953, based on work done in the 1940s), pulls together the work of a larger group of people who were using symbolic events produced by members of a culture to find patterns of that culture.

"Films," she wrote, "being group products, have proved to be more immediately useful for the analysis of culture than have individual literary works." In this book she included the first set of systematic analyses of films by a group concerned with looking for cultural forms and the patterns evidenced in them. This work provided a cornerstone on which almost all the content analysis of our current mass media rests.[2] The development of the Cultural Indicators program (Gerbner 1972; Gerbner and Gross 1976) and the ongoing analysis of mass media, and particularly of TV content, are the fruits, it seems to me, of one direction developed from the notion that the photograph, in still or motion picture form, can be a record of culture in its own right, to be studied for its own patterns within specific cultural contexts.

The term *visual anthropology*, coined after World War II, became associated with conceptualizations keyed to using cameras to make records about culture. The term did not seem to connote studies that led us to ask what we could learn about a culture by studying what the members of a society made pictures of, how they made them, and in what contexts they made and looked at them.

The ideas of modes of symbolic communication designed to articulate a variety of symbolic worlds is not new to social science. Cassirer, Whorf, and many others discussed the idea that symbols and symbol systems, language, myth, stories, and conversation, as well as poems, sonatas, plays, films, murals, and novels, create a multiplicity of worlds.

Nelson Goodman addressed himself to this line of speculation at a meeting commemorating the one-hundredth anniversary of Cassirer's birth. He asks a set of questions that I would like to use to discuss some of the current issues we face in an ethnography of visual communication. He asks, "In just what sense are there many worlds? What distinguishes genuine from spurious worlds? What are worlds made of? How are they made, and what roles do symbols play in their making?" (1975:57). I think that it is only recently that we have been able to apply these questions to an endeavor we call anthropology, to a mode I call pictorial-visual, and to a concept that has come to be called communication. It was Margaret Mead who helped, not only by her work but by her teaching and her encouragement of the work of others, to integrate those three concepts: anthropology, communication, and the visual-pictorial mode.

When in 1963 (Worth 1964) I began to point out that films and photographs made by such diverse groups as students in college, people in their homes, or mental patients in hospitals could be looked at as ways in which these different people structured *their* worlds, rather than as "true images" of *the* world, I thought I was merely bringing a truism about drawing and painting up to date. Most people who talked knowledgeably about pictures in 1963 accepted the fact that Picasso drew the way he did because he meant to structure his pictures that way, not because he could not draw like Norman Rockwell or even the way he himself drew in other periods. True, Roman Jakobson in 1950 pointed out that most people wanted pictures to look like a Norman Rockwell—what we now call photographic or snapshot realism—and were disturbed by abstract painting. Jakobson ascribed this both to the fact that most people were ignorant about the conventions of painting and to the strength of conventions about pictures—when they were known. He, himself, it seems, tended to believe that the "natural" way to know pictures was to know what they represented; that to draw abstractly, or in non-representational or non-Western patterns, was somehow to act unnaturally. Interestingly, it was the early Russian filmmakers and film theorists—Eisenstein, Dovzhenko, and Pudovkin—who, following the Russian formalist linguistic theories, first pointed out that films structured reality just as speech did; that patterns of images, like patterns of sounds, were worthy of study. But so strong was the myth of photographic reality that even a Roman Jakobson could feel that representation was the natural way to make pictures.

For many leading social scientists today, as well as for our students, visual anthropology means taking photographs, photo records, movies, ethnographic movies, and film footage—all for research. These labels carry a descending aura of science about them. Film footage, unorganized but uncut, is considered the most scientific, and therefore the truest evidence, because it captures "real behavior," presumably untouched by human eye or brain—a pure record. An ethnographic movie or a documentary movie is the least scientific, not only touched, but sometimes, it seems, tainted by human consciousness and often damaged as a scientific document by something called "art." As recently as last year, the chairman of the Department of Sociology at Columbia University wrote in the *New York Times*, in shock, that a documentary film about the Yerkes Primate Laboratory expressed the filmmaker's biased view of the subject, still naively stating that he expects something called a neutral, unbiased, objective view in a film shown on television. The director of the laboratory, who had given permission to the filmmaker to make a movie to be shown on television, expressed anger that the film did not portray the "truth" about the laboratory. He, too, evinced shocked dismay that the filmmaker presented his own personal view of what he observed in the laboratory—that the act of making a movie allowed such a "distortion."

There is no point, however, in taking a position that if film is not "objective truth," there is no use to film. Many ethnologists have provided us with stills and motion pictures which they and others have used to articulate some of the most important statements about culture made in recent years. I am arguing that there is great value in visually recorded data about behavior and culture, so long as we know what it is that we recorded, so long as we are aware of how and by what rules we chose our subject matter, and so long as we are aware of, and make explicit, how we organized the various units of film from which we will do our analysis.

Let us return to Cassirer and Goodman's concept that symbolic events produce different works and different worlds. Faced with such a concept, and, most specifically, with the fact that pieces of film, no matter how made, are patterned constructions, structured, at best, by a trained mind, the truth-seeker through film becomes confused, dogmatic, and angry. It is hard enough for some people to believe that an analysis is a construction, a structuring of reality. Most of us simply do not want to face the fact that what we loosely call primary photographic data is also a structured event. A photograph, just like any picture, is constrained both by who made it and how it is made, as well as by what it is a picture of. It should be obvious that just as pictures are not simple mirrors of what is out there, neither are they artifacts which have no relation whatsoever to what they are pictures of. The ethnographic photographer is free to take a picture of anything his system allows him to photograph, but he is also constrained by the fact that he must point the camera at some objects in the world "out there." These things out there also constrain what the picture will be like. While "out there" does not determine what the photograph will look like, it is obviously not irrelevant. In one sense, we want as many different worlds as possible and in another, the fact that symbols and signs can best be used to construct different worlds poses almost insoluble scientific problems. In order to distinguish genuine from spurious worlds, we slip into the belief that cameras record reality, that reality is true, and that film recordings are therefore "truth."

This fantasy about symbols suffers from the error of imposing logiclike or logical sounding rules upon a domain that is governed by a set of rules that may not be like those of logic. For example, one basic convention of logic states that a true conclusion cannot be drawn from a false premise. Researchers who want to use film as a record of behavior want it to be the case that from a true premise—a picture or photograph—one cannot draw a false conclusion; that is, that from "true" films one cannot get "false" data. One introductory lecture in logic should be enough to make any student see that this is not the case. Unfortunately, false conclusions can be drawn from anything, and getting the "truth" on film, even if it were possible, will not guarantee the subsequent analysis or the conclusions drawn from it.

Suppose we agree that pictures and films can be used as illustrations of pattern—of how films themselves are structured, as well as of how people and their behavior in films are structured. Suppose we agree that symbolic events produce symbolic worlds, and that these worlds are not (for the moment) to be thought of as either true or false, but rather as communicative articulations. Suppose we think of a film, whether it be footage without editing or footage after editing, as the way the maker of the film structures the world that he or she presents to us. Our job as viewers, then, is first to determine what he means by the film he shows us. A mere recording, without conscious selection, emphasis, and instruction by the filmmaker, is more often confusing than illuminating. The viewer of such a recording "knows" that an inanimate camera did not expose the film and decide what to shoot and how to shoot it. If the film does not instruct us how to interpret it, or if it is not constructed in a way that allows us to use conventional techniques for interpretation in that medium, we most often ignore the film or treat it as an annoyance. Ray Birdwhistell, with whom I have watched too few films, has often said to me, "I can't stand watching most so-called ethnographic movies. The man who made it won't tell me what he's doing. I'd rather look at behavior as it occurs and not have to spend all my time trying to guess how, when, and for what reasons a filmmaker made a movie of it."

Seven years ago, again led by Margaret Mead, a group of researchers interested in both records about culture and records of culture met and decided that our concerns could best be clarified by founding a new organization, with its own journal. Margaret Mead helped us to set up the Society for the Anthropology of Visual Communication (as part of the American Anthropology Association) and the National Anthropological Film Center at the Smithsonian Institution.

The kinds of problems that our members study include all the ones that I have mentioned (for there are indeed still not enough systematic records about the cultures of the world that can be used to illustrate patterns of culture), as well as the newer ones I will be talking about in a moment.

In developing a history of the shift from visual anthropology to the anthropology of visual communication, and in trying to understand Margaret Mead's role in this development, it is most important to understand that the study of culture is not accomplished by pitting symbolic worlds against one another. Those of us who are involved in using photographs and films as new technologies through which we can record cultural artifacts and events, and those of us who are involved in studying how pictures are put together to make statements about this world, are equally concerned with learning how this particular symbol form, the picture, can be of use in the study of culture. We include scholars such as Richard Sorenson (1976) and Jay Ruby (1975,1976), who are struggling to delineate theories of the photograph as evidence, as well as those who are following up on the work that John Adair and I (Worth and Adair 1972) did when we gave movie cameras to Navajo Indians to see how their patterns of structuring differed from or resembled ours. Most recently, *Studies in the Anthropology of Visual Comunication* devoted a complete issue to a study by Erving Goffman of values and social attitudes about gender that can be derived from an analysis of some 500 advertising photographs (Goffman 1976).

Some of us are arguing that it is as silly to ask whether a film is true or false as it is to ask whether a grammar is true or false, or whether a performance of a Bach sonata or a Beatles song is true or false. The confusion about the use of pictures, in social science particularly, arises out of the fact that although symbol systems are designed to articulate many worlds, our way of thinking about such systems allows us, even compels us in certain contexts, to ask, "Are you trying to tell us that all symbolic worlds are equally true, equally correct, equally right in their portrayal of the 'real world'?"

One can indeed ask if a particular grammar is a useful description of how people talk. One can ask whether that sonata was written by Bach or whether that was a Beatles song. If the notion of a grammar is understood to be an articulation, a statement about how people talk, one can ask in what ways it corresponds to how people do talk. But this requires that we conceive of a grammar, a performance, or a film as a statement or a description of and about something. It requires that we understand that the grammar or the film is *not a copy of the world out there but someone's statement about the world.*

Acknowledging this, some of our younger colleagues are beginning to study such things as how home movies are made as a social event, as well as what they mean as a semantic event. We are looking, as Chalfen (1975) has done, into how home movies and photograph albums are displayed and exhibited, to whom, and for what social purposes. Ruby has begun to study the patterns apparent in the photographs that most anthropologists make in the field. Here he finds that in most cases they are indistinguishable from those made by journalists. That while anthropologists' written ethnographies do in fact differ from journalists' reports or travelers' letters home, their photographs do not differ from the kind that journalists take. For the most part, anthropologists and (as Howard Becker [1974] has shown sociologists are professional scientists—only when they are employing words. When it comes to the visual mode of articulation and data-gathering, most produce snapshots, documentary films, good (or bad) home movies, or "artistic" works. It is forty years since Bateson and Mead took their photographs in Bali, and, sad to say, in that forty-year period there have not been many social scientists who have been trained in what they developed.

The framework of the anthropology of visual communication suggests that symbolic worlds are patterned and amenable to being studied in a larger framework than pictures. Primarily, this framework helps us to look at pictures as that aspect of culture called communication. It suggests that we treat pictures as statements, articulated by artists, informants, scientists, housewives, and even movie and TV producers. We can ask what the articulator meant, and then we can ask whether our interpretation of what was meant is good, bad, beautiful, ugly, and so on. But by asking whether our interpretation of what was meant is true, we are, I am afraid, merely asking whether we guessed right. What we should be trying to understand is how, and why, and in what context, a particular articulator structured his particular statement about the world.

Treating film (the camera and celluloid) as a copy of the world, rather than as materials with which to make statements about the world, forces us into the impossible position of asking whether performance is true. Understanding that photographs and films are statements, rather than copies or reflections, enables us to look explicitly, as some of us are now doing, at the various ways we have developed of picturing the world.[3]

The parameters along which we deal with statements are many. Anthropology is in some sense a set of questions about human behavior. Ethnography is in some sense a method by which certain kinds of questions can be answered. By considering pictures and all behavior in the visual mode as possible communication acts, and by understanding that these acts can produce only statements or assertions about the world, rather than copies of it, we are enabled to consider the kinds of anthropology we want to do about the visual pictorial forms that we can and do use. In this kind of anthropology, we want to consider both how the photograph and the film can be used as evidence by the scientist and how people actually have used them, as evidence, as documents, as entertainment, and as art.

It is only within this framework that we are able to return to Pat Loud's questions with which I opened this discussion. Margaret Mead actually did influence that "show," just as she did influence this paper. Craig Gilbert, the producer of "An American Family," was previously the producer of another show, "Margaret Mead's New Gunea Journal." Gilbert spent a great deal of time talking with Dr. Mead about films and about culture, while he accompanied her on her return trip to some of the places she had studied in the past. He learned from her that one American family, well observed, might reveal or, in her words, "illustrate" a pattern about American families. The patterns that he observed and the way they are structured are his, and his cameramen's and editors'. The idea of trying to present them on film was learned from Dr. Mead.

Pat Loud said it "didn't work," that when she saw the film of her family, she felt herself "shrinking in defense." She felt that she had been "ground through the media machine" and "treated as an object." Then she said she would do it again if it did what the producer said it would do. Craig Gilbert had told her that by showing one family he could show a pattern that might be true of many American families.

We know now that it was not the editing that prevented the programs from "working." We have tried to reedit some of that footage. We have invited Mrs. Loud to do it herself. It seems that it cannot be done so that the film does not look as if it were produced as a drama or a soap opera for TV. Because it is on TV. And TV does not present the truth any more than film does, or than film editors do. It presents, we now know, a structured version of what someone saw, presented in a context—television—of drama, soap opera, sporting events, "news," and commercials. We have learned how to interpret what we see on TV. If we were to study that footage in other ways and not show it on television, we might find patterns that would illustrate other structures, other worlds.

Learning how to study something as complex as a twelve-hour film put together from 200 hours of film based on 400 hours of observation is part of the study we are now calling the anthropology of visual communication.

There are now heated controversies about whether Mrs. Loud and her family were fooled, whether (leaving television aside) sociologists and anthropologists have the right to photograph real people for their studies. Again, in 1936, and reported as early as the second page of *Growth and Culture*, Dr. Mead faced this question. She wrote, "I have used real names throughout. The people knew we were studying and photographing their children; indeed, they often helped set the stage for an afternoon's photography. Very cautiously, but quite definitely, they gave us permission to live among them and there is no need to blur their contribution by disguise or subterfuge." Adair and I followed this advice in our own work among the Navajo, first getting their permission and then acknowledging their great contribution. They were in their own films, and they wanted to be seen. We can tell what would have happened had the press and assembled academics called them primitive, selfish, cruel. As we have described in our book about this project, they themselves did not think of their films as the truth about Navajos. Their films were true about, as one of them put it, "how you tell a story." Those of us interested in the anthropology of visual communication are trying to find ways to study how people can and do depict mankind, themselves, and others in all their diversity.

In 1967, I returned from the field with 12,000 feet, 480,000 single frames, of exposed film, and seven movies made by Navajo Indians. I was looking for patterns, but I was overwhelmed (as so many researchers are when they return from the field) by the masses of observations and possible data I had collected. The patterns were far from clear in my mind. I was tired. Dr. Mead asked me to show some of the films and talk about my research to her class. I did. The next day after breakfast, she quietly set up the projector, pulled up her typewriter, and asked me to start going over the footage with her. I had worked with this material for over a year. Margaret Mead began to teach me how to find patterns in it. When I finally said something like, "I know that, why do we have to keep going over it?" she replied somewhat tartly, "Sol, you begin with intuition, but you can't rest your case upon it. You must build upon it and make clear to others the patterns that seem clear to you."

This paper is my continued attempt to follow that advice. Doing the anthropology of visual communication is an attempt by a large group of students of communication and anthropology to find methods and theories by which they too can make clear the patterns that they discover and create.

Notes

[1] This paper was presented at a symposium honoring the work of Margaret Mead at the annual meeting of the American Association for the Advancement of Science (1976) and published in *Studies in Visual Communication* 6:1 (1980): 15-22—Ed.

[2] For a more detailed exposition of the relation of content analysis to the analysis of culture through pictures, see introduction to Erving Goffman's "Gender Advertisements" in *Studies in the Anthropology of Visual Communication* (Worth 1976: 65-68).

[3] For a specific study of how advertisements picture the world, see Goffman 1976.

References

Becker, Howard. 1974 "Photography and Sociology" from *Studies in the Anthropology of Visual Communication* 1:3-26.

Chalfen, Richard. 1975 "Cinema Naivete: A Study of Home Moviemaking as Visual Communication" *from SAVC* 2:87-103.

Collier, John Jr. 1967 *Visual Anthropology* New York: Holt Rinehart and Winston.

Gerbner, George. 1972 "Communication and Social Environment" from *Scientific American* 227:153-60.

____________, and Gross, Larry. 1976 "Living with Television: the Violence Profile" from *Journal of Communication* 26:173-99.

Goffman, Erving. 1976 "Gender Advertisements" from *SAVC* 3:69-154.

Goodman, Nelson. 1975 "Words, Works, Worlds" from *Erkenntnis* 9:57-73.

Loud, Pat and Johnson, Nora. 1974 *A Woman's Story* New York: Bantam Books.

Mead, Margaret and Metraux, Rhoda, Eds. 1953. *The Study of Culture at a Distance* Chicago: University of Chicago Press.

____________, and MacGregor, Frances C. 1951 *Growth and Culture: A Photographic Study of Balinese Children* New York: Putman Press.

Ruby, Jay. 1975 "Is an Ethnographic Film a Filmic Ethnography?" from *SAVC* 2:104-11.
———. 1976 "Anthropology and Film: The Social Science Implications of Regarding Film as Communication" *Quaterly Review of Film Studies* 1:436-45.
Sorenson, Richard E. 1976 *The Edge of the Forest: Land, Childhood and Change in a New Guinea Proto-Agricultural Society* Washington DC: Smithsonian Institution Press.
Worth, Sol. 1964 "Public Administration and the Documentary Film" from *Perspectives in Administration (Journal of the Municipal Association for Management and Administration, City of New York.)* 1:19-25
———, and Adair, John. 1972 *Through Navajo Eyes: An Exploration in Film Communication and Anthropology* Bloomington: Indiana University Press.

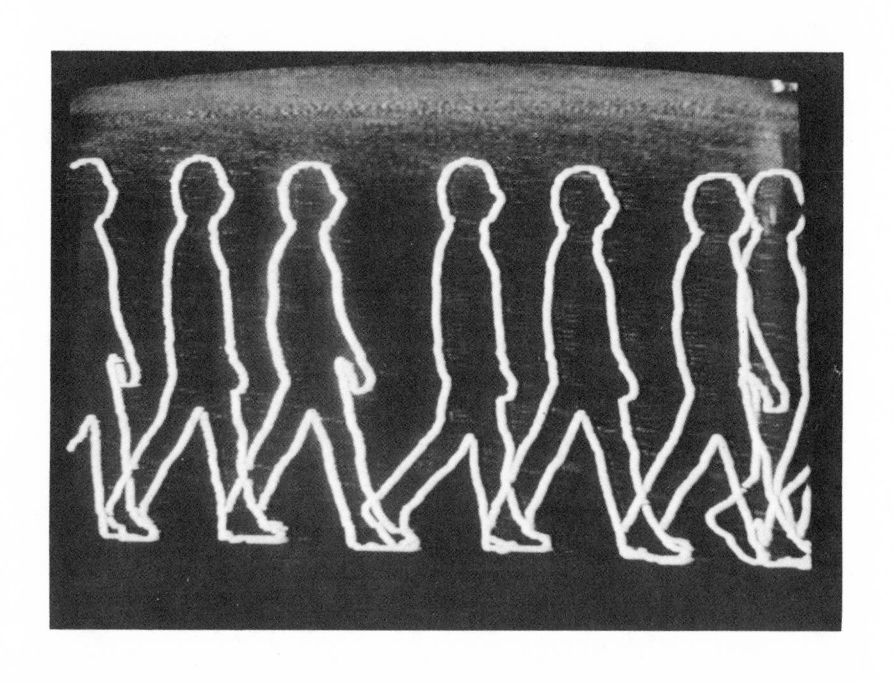

CRACKING THE CODES OF TELEVISION: THE CHILD AS ANTHROPOLOGIST

Howard Gardner
(With Leona Jaglom)

THE ANTHROPOLOGIST occupies a unique position in our society. While most of us have little opportunity to visit exotic lands, such ventures constitute the anthropologist's central mission. It is a daunting one. Relying primarily on common sense and general knowledge of human nature, he has to describe and eventually to construct, almost singlehandedly models of various aspects of an entire society—its language, its kinship structure, its values, and its beliefs. He must continue to test and revise his formulations as necessary, until he feels relatively secure in his characterizations. Even with the help of articulate informants, he is unlikely to grasp the culture in all its particulars. Indeed, if his description is even approximately right, his ethnography will be considered a success.

Though our daily routine may seem far removed from such an existence, most of us have been anthropologists early in our lives. For in being placed in front of a television set and being asked, in effect, to make sense of the innumberable fleeting images it presents, the young child of two, four, or eight years of age is a kind of anthropologist.

Indeed in many ways the job confronted by our young anthropologist as he steadily eyes a Sony portable set is even more challenging. He has spent but a few years on the planet. He must make sense of diverse and seemingly incommensurate forms of reality, including the technical apparatus placed in front of him, the various slices of reality it serves up, the complicated panoply of commercials, features, news documentaries, entertainment, specials, and test patterns which constitute daily video fare. One might say that he has to decode or unravel a number of worlds—the world of television as a whole, the world presented on each channel, the world represented by each kind of program, each particular program, and each episode and scene.

To complicate matters further, the child must learn the visual language used by television (for example, closeups, instant replays, montages). And he must decode rules that determine the operation of commercial and public television, the relationship among the various channels, the motivations leading to the production of commercials and shows, the status of live shows, recorded shows, original productions, and reruns. These tasks might intimidate even an accomplished Oxonian ethnographer.

But amazingly, the average child, working largely on his own, succeeds in making sense of these worlds in remarkably short order. By the age of five, most of these understandings and discriminations have already been accomplished. In fact, so speedy and untutored is this process that it recalls the preternatural facility shown by children everywhere in mastering language and other symbol systems of their society.

All the same, the child's conquest of television is neither total nor totally even. Just as sophisticated aspects of language may elude the young school child, certain aspects of television, both technical and programming, remain opaque even to much older and more accomplished youngsters. And confusion attends such conventions as canned laughter, such distinctions as documentaries versus fictionalized history, such realities as the purpose of commercials. It is here that skilled informants—such as parents, teachers, or peers possessing higher levels of sophistication—may prove essential. An especially delicate matter involves the proper situating of television within the child's ongoing life stream, giving it a place that is neither too exalted nor too inconsequential. Like the anthropologist, the child must steer between excessive ethnocentrism and "going native."

At Harvard's Project Zero my colleagues and I have been studying the first years of television viewing. We have been observing a small group of first-born, primarily middle-class youngsters between the ages of two and five as they watch television. Donning the roles of anthropologists ourselves, we have observed the children as they ingest in turn a steady diet of "Sesame Street," "Mister Rogers," the assorted daily mix of cartoons, commercials, soap operas, situations comedies, and the occasional specials that punctuate the television season. Watching children watching television, we have documented the tremendously rapid mastery of basic video competence and detected as well some lingering problems in "reading" the worlds transmitted on television.

At the outset every child must achieve two fundamental understandings about television. First of all, the child must come to understand the nature and limitations of the *physical medium* of the television set. To the extent that a one- or two-year-old child attends to television (and most of them at least sample the wares of video), these youngsters accept the material presented on television as a natural part of everyday life and regard the television set as a member of the family. There is essentially no appreciation of television as a distanced medium, one whose content represents rather than constitutes the daily flux of experience. Thus, at twenty-five months one of our subjects, whom we call Johnny, sees a broken egg on television and runs to fetch a paper towel to clean up the egg. Donny, just two months older, is afraid that the Abominable Snowman on television will want to invade his room. Sally, at the same age, spanks a television personage and then kisses her, and at age three my son Jerry became terribly fearful when, while sitting next to me, he spied me appearing on a television talk show as well. Not until the child is four or five does he understand that what is presented on television exists in a world apart from his immediate life space.

A second and equally fundamental challenge facing the child during his first years of viewing is an appreciation of the *narrative nature* of much of television. For the one- or two-year-old, television presents an army of isolated images that bear no connection to one another. Any image could appear at any time: no necessary order seems to govern beginnings of programs, commercials, or station announcements. By the age of three or four, the child begins to sense that television presents narratives that may be interrupted from time to time by commercials. However, even when the narrative purpose of television has been grasped, the understanding of individual narratives may remain extremely meager.

In an effort to trace the beginning of narrative competence, we studied several youngsters as they observed the "Sesame Street" bit, "The Boy Who Cried Monster" (patterned after "The Boy Who Cried Wolf"). Over a period of three years, the children watched this bit dozens of times with undiminished interest, tinged during the latter years with fright at the sight of the scary cookie monster. Comprehension of the bit surely increased over this period. Children came to realize that the vignette did not depict merely a voracious monster who deprived a boy of cookies: by the age of four or five there was incipient awareness that the villagers played a role in the story and that the boy had antagonized them with false alarms. Yet we found that even the four- or five-year-old child had only a partial understanding of this still simple and familiar fairy tale—failing, for example, to grasp that this particular monster was actually a friendly and grateful devourer of cookies.

Any sophisticated relationship to television rests on these fundamental understandings: the world presented on television exists independently of the actions and thoughts of the television viewer; the bits presented on television programs are ordinarily related in a sequential order. Despite the various misinterpretations just described, most children do appreciate these concepts by the conclusion of the preschool period. It is worth noting that such understandings of how to "read" symbolic products emerge more readily in the case of books than of television. As physical entities, books seem less "true to life" than television and so are less likely to be confused with daily reality. Moreover, both the slower pace of books and the ways in which their contents are customarily shared between parent and child, encourage mastery. Still, whatever confusions may obtain with respect to the apparatus of television and its fundamentally narrative quality have largely been dispelled by the time children reach five years of age.

Three remaining puzzles perplex the youthful television viewer. The first is the need to relate what is presented on television to what is available in the rest of one's daily experiences. The child must arrive at the following conclusion: the world of television is not exactly the same as his everyday world, but at the same time it is not totally alien from that world either. As we have come to phrase it, the child must be able to appropriately construct that *membrane* which stands at the interface between the worlds of television and world of daily life.

Recall that the child at the age of one and a half accepts television as an unquestioned part of his daily experience. By the age of about two, while acknowledging the materials encountered on television may not be identical to ones from his own world, he remains mesmerized by similarity. Thus, at twenty-eight months Sally sees a frog on television and immediately runs to get frog puppet and plays with it. Vocal comments at this time also simply underline the similarities between materials or events presented on television and counterparts from the child's own existence: "I go McDonalds," "Get Lego," "Let's fly plane."

Having been hitherto dominated by identities and similarities, the three-year-old child effects a transition to the noting of differences between the televised and real world. Sally at forty months sees a stuffed bear and comments sagely, "Polar bears have this many feet? I never saw that on 'Sesame Street.'" This detection of differences may eventually assume such ferocity that the child comes to deny any relationship between what is presented on television and what is encountered in the real world. Thus Donny at fifty-seven months questions whether sneakers seen on television could be like sneakers in his own house; Jonny at fifty-nine months denies that he could be on television even though he has recently seen himself in that medium; and Sally, nearly five, claims that real people could not be dressed as Wonder Woman because "they're not real when they're on television."

In the course of the first few years of television viewing, then, the child moves from a time when there is no membrane to a phase when the membrane between these two worlds is virtually impermeable—a sort of Berlin Wall separating television from daily life. By the age of five, however, the membrane has achieved a more appropriate, semi-permeable status. The child regularly points out both the similarities *and* the differences between television and the real world. Seeing Kermit the Frog on television, Sally declares, "I have a puppet like Kermit too, but mine doesn't talk."

The emerging acknowledgement of semipermeability reflects increasing awareness of the symbolic or representational nature of television fare. For example, seeing a fire on a news program, Charlie, age four and a half, realizes with genuine surprise (and cognitive pleasure) that it is the same fire which had been discussed at his school earlier that day. Newly aware that television may sometimes (but only sometimes) represent a specific aspect of his own daily experience, the child becomes gradually more open-minded about the connections between the domains.

As part of our investigation of the relationship between television and real life, we wondered which domain serves as a primary point of reference for the child. Hence we searched for references to daily experience while the child was watching television, and for the references to television during the child's non-viewing hours. Somewhat to our surprise (and to our considerable relief), we found that the world of mundane experiences typically functions as a backdrop against which television is viewed. Children often remark that something on television resembles experiences in the real world: they much less frequently characterize daily experience in terms of the extent to which it resembles television. At least in our small population, there was little indication that television serves as a "real basis" against which to evaluate daily experience.

A second major puzzle facing the young television viewer is the relation among the multitude of programs presented on a typical day. The solving of this task involves recognizing various shows as well as diverse kinds of programs and charting their relationship to one another. Entailed as well is an appreciation of the schedule that governs programs and the role of "central characters" in the definition of a program. We were interested in the kinds of program distinctions children could initially make, and in the factors that motivate these and further delineations.

In general, the child's discriminations among television fare echo the distinctions he is effecting between the televised and real world. In the beginning the membranes obtaining among kinds of shows are totally permeable. Any character or any segment can appear, willy-nilly, at any time or in any context. Indeed, the young child attributes to himself magical powers: if he wants a bit to appear, it simply shall. To the extent that shows have any definition at all, they are built primarily around their lead characters.

By the age of two and a half or so, the child does make certain very broad distinctions among television programs. For better or worse, the first reliable category to emerge is the commercial. At a time when the child has no understanding of the purpose of commercials, he is already able to identify and label those short segments as different from the rest of television fare. We believe that the brevity, clarity, and highly vivid character of commercials contribute to their recognizability. In contrast, though our children watch the television show "Sesame Street" much more frequently than they watch any commercials, they do not recognize it as a show. Possibly the "magazine format" of "Sesame Street" renders it much less identifiable as a specific program or program type.

By the age of three or three and a half the child can recognize and identify cartoons: he also comes to appreciate that the various magazine segments juxtaposed together all fall under the single rubric of "Sesame Street." Children prove to be instructively sensitive to a relatively rare feature of a television diet—the preview for a forthcoming special. We speculate that previews—a kind of commercial—are especially recognizable because of the great interest that these specials hold for children, and also because they serve as hypothese of future events which are confirmed when the program itself eventually appears on television.

Certain telltale cues may signal specific programs and program types, but they do not in themselves allow for complete mastery of program distinctions. Thus unless children also appreciate the temporal organization of television—when shows begin and end, how flexibly specific characters and bits may appear across time slots, the nature and timing of commercials and station breaks—they will remain confused. In fact, such organizational aspects pose profound problems at the beginning; at age two or two and a half children have little recognition of what should appear at a given time, confuse commercials with programs and beginnings with ends, and stubbornly insist that characters will return even after a show has been completed. By the age of three or four, children do recognize the most familiar beginnings and endings, have a sense of schedules for their favorite television programs, and acknowledge that when a character fades with the end of the show "he won't be back again 'til tomorrow."

Children's understanding of television fare takes a decisive leap forward in the period between age four and five. At last children have a firm sense of the beginnings and endings of shows, the times at which the shows and characters appear, and the mutually exclusive relationship between regular shows, advertisements, and specials. Certain other program distinctions continue to pose problems. Even though (or perhaps because) from the age of two children dislike "the news," they still have difficulty reliably identifying segments taken from news programs. They also fail to honor the well-worn programmer's distinction between "children's programs" and "programs for grownups." In fact, they make a charming confusion: children's programs are simply programs watched by children, whether they are opera, soap opera, news, sports, or a segment of the "Beverly Hillbillies."

This confusion highlights a revealing aspect of children's television viewing. While adults readily classify programs in terms of the intent of the producer and the designated target audience, children have great difficulty appreciating these abstract modes of classification. They think instead in terms of particular shows or even particular episodes and resist grouping these together as instances of situation comedies, dramas, documentaries, and the like. To the extent that they ever offer overriding generalizations, they will classify on the bases of the affects they experience: "funny shows," shows that are "frightening," and "boring programs."

We again encounter the "membrane effect" as the child ponders the role of the central character in a program. Initially, as we have seen, children assume that characters can appear at any place and at any time. Later on, during the middle preschool years, children come to associate characters exclusively with particular programs: they refuse to acknowledge that a character associated with one program can appear on another. And so if Mr. Rogers happens to appear on a talk show, that talk show becomes "Mr Rogers"; or if Flip Wilson appears on "Hollywood Squares," it is then the "Flip Wilson Program." Eventually the membrane softens. The child of seven or eight does concede that an individual character is somewhat mobile in the land of television: even though he has a primary home, he may be allowed to visit other houses on the block.

The final and perhaps most formidable puzzle confronting the young television viewer is the status of, and relation among, the multiple levels of reality and fantasy within the variegated fare presented. In artistic media, levels of reality and fantasy are typically mixed: television is perhaps the medium that serves the most notorious and dizzying ensemble of levels of reality. While continuing to remain confused about this melange of "reality with tiers," children nonetheless do make significant progress during the early years of their lives in effecting some preliminary sorting.

Initially the child is likely to consider everything presented on television—live films, cartoons, photographs, fantasy, or news—as equivalently real. Reality is the background against which fantasy or nonreality stands out. The first distinguishing markers surround abnormal or superhuman figures like monsters, wizards, and freaks. Children sense their divergent reality status. But frequently the child's own wishes get in the way, as he declares that he wants to be (and could be) like the Hulk, or Superman, a prince or a king. And even when the child purports to recognize the lines between reality and fantasy, this understanding may cloak amusing ignorance. Consider the child who said, "I know Big Bird isn't real, that's just a costume. There's just a plain bird inside."

If the potency of the character provides one important clue to irreality, the medium of presentation is another. The closer to photographic reality, the more likely a character will be considered "real," and conversely, cartoon figures, figures executing impossible actions, or figures presented in strange iconographies are much less likely to be judged as real.

But to transcend this very preliminary and primitive sorting, the child must master two long-term agendas. First of all he must become alert to the various uses and tricks to which the medium can be put, to the effects that directors and editors can conjure up. Only then can the child peer through surface indications of what is real and what is not and make a cool assessment of the realistic status of a segment.

The second requirement is to go beyond the simple duality—possible-impossible, real-fantastic—and to render an assessment of plausibility. Among the innumerable scenes and situations presented on television, certain bear a much closer relation than others to events in the world of regular experience. Yet the capacity to invoke plausibility presupposes the ability to weigh numerous factors simultaneously and to render a probabilistic judgment—calculations far beyond the ken of the young child. And so children must fall back repeatedly on simple formulae and single markers, which often yield erroneous assessments.

To override these initial canons of classifications proves challenging. Only during the school years do children become alert to the fact that one individual may look real but behave in fantastic ways (Gilligan, for example), while another may be animated yet undergo psychologically authentic experiences (Charlie Brown, for example). And only at that time does a child become even dimly aware that certain kinds of programs (such as historical fiction or documentaries) may entail a peculiar blend of veridicality and editorial judgment.

How then does our young television viewer stack up as an anthropologist? On nearly any criterion, he performs extraordinarily well. Some aspects of television—for example, its narrative nature—are solved early on. More challenging aspects—the schedule of television, principal distinctions among programs—are essentially understood by the time the child enters school. Given the child's meager experience and relative lack of help from informants, his achievement is spectacular.

Nonetheless, as exemplified by children's continuing difficulties in sorting out various levels of reality, certain aspects of television viewing remain problematic. Separating the facts from the editorials in a news broadcast, the amount of staging in a game show, or the extent to which various celebrities on a talk show are "being themselves," proved extremely difficult matters.

Another problem of enduring difficulty is to determine the accuracy with which the world is portrayed on television. While our young children consider the real world as the basis against which to assess television, the steadily accruing experience that children have with the various stereotypes fostered by television may ultimately tip the balance. Children may know so much more about law courts, operating rooms, or racial groups from their viewing experience than from their daily lives that they will inevitably attribute a high reality status to these video stereotypes and come to evaluate their occasional actual experiences against this idealized reality. As Jerzy Kosinski remarks in *Being There*, "In this country, when we dream of reality, television wakes us" (p. 89).

More complex narrative materials on television pose continuing challenges. While even the young child frequently gets the basic point of the plot, there is mounting evidence that important aspects of television programming continue to pose difficulties for children and even adults. In addition to the work in our own laboratory, we can invoke the recent report of Gavriel Salomon that American school children cannot answer even simple facts about television programs viewed on the previous day, and the findings of Purdue's Jacob Jacoby that a very high proportion of television viewers misunderstand at least part of what they see on short thirty-second segments of programming or public-service commercials.

Some of these aspects of television would prove difficult for any individual. Assessing the plausibility of a deliberately ambiguous film like "The Autobiography of Miss Jane Pittman" or the televised series "An American Family" would challenge even a philosopher. Even if these deliberately provocative examples are bracketed, it seems clear to me that children can be helped in dealing with questions of ambiguity, authenticity, and stereotypy. This is where parents, older siblings, friends, and educators can make a contribution; indeed, the people working in television can themselves perform a public service. If the brains that are customarily put to work in fashioning effective commercials were instead (or in addition) marshaled to clarify the nature of challenging materials presented on television, progress could be made in sealing various knowledge gaps confronted by the school-age child and by many adults as well. Dorothy and Jerome Singer of Yale University have made promising efforts in this direction through the devising of segments that take viewers "behind the scenes" of television.

Television can be mastered. Indeed, human beings seem extraordinarily well equipped to make initial sense of this complex medium. But even as language has its arcane and complex aspects, involving abstract argument, rhetoric, or figurative expression, so too can television present excessively challenging material and may, if somewhat inadvertently, foster obfuscation rather than clarity. Equipped with a good handbook, a skilled informant, and his own intelligence, the competent anthropologist eventually learns even the most difficult native tongue. Given analogous help, children can equally master the full range of television fare and will certainly enjoy it better. They may even succeed in becoming superior producers of television material. Quite possibly those whose "native tongue" is television will eventually become its greatest poets.

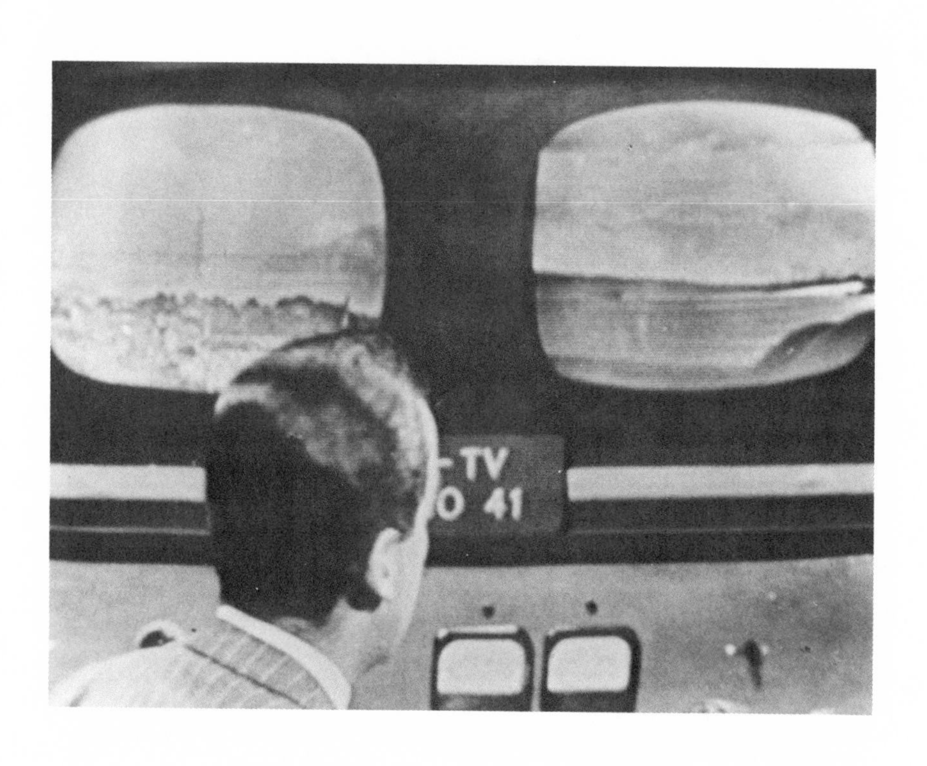
TV
41

INTERACTIVE TELEVISION

John Carey and Pat Quarles

Introduction

IN THE MID 1970S, there was a flurry of excitement in the commercial telecommunications world as well as education and government about the potential of interactive television. A "wired nation" appeared to be just around the corner and with it came a promise of a technological promised land in which every home would have a two-way link to virtually unlimited information and entertainment. One of the projects begun during this period was based in Reading, Pennsylvania: interactive cable television for senior citizens. Created with funding from the National Science Foundation and implemented by a consortium of New York University and Reading groups, the project was field tested in 1975 and formally launched in 1976. Berks Community Television (BCTV) as it was later to be called, is now approaching ten years of operation. Curiously, it is one of the few interactive cable television projects launched in the 1970s that still operates. Indeed, the two-way cable service has expanded over the years to include many other user groups and origination sites within Pennsylvania's Berks County. Why has the Reading interactive cable service succeeded while many others have not? Further, why has this apparently successful community channel not been imitated by others? To tackle these questions, it is useful to trace the history of interactive television, describe how the Reading system was created and review its current status nearly a decade later. Emerging from this assessment is a picture of a social communication innovation in which technology played an important but secondary role.

Interactive Television

Interactive television is not a single technology or service, but a family of diverse systems and applications which trace their history to the very beginning of television. Among the earliest experimental television projects in the 1920s, was two-way video communication, In the 1950s, a few rather primitive interactive formats were introduced to commercial television. A children's program in New York, *Winky Dink And You*, created simple interaction through the use of a plastic screen that viewers attached to the regular TV screen. In the programs, a cartoon character Winky Dink would encounter a series of problems. For example, chased by a tiger Winky Dink would run to the edge of a chasm. Children watching in their homes were then exhorted to help Winky Dink escape by drawing a bridge with crayons, on their plastic screen, over the chasm. One problem with this format was the participation of kids who failed to purchase the special plastic screen and simply drew with crayons on the glass TV screen. Another well known interactive format was imployed by Edward R. Murrow in his interview program *Person To Person*. Here, a guest in his or her home was interviewed by Murrow, who remained in the studio. Often, there was a considerable distance between the two locations. While the program enjoyed commercial success, the interactive format was very crude. In fact, the guests could not see Murrow though they often pretended to see him.

Interactive television received a major promotional boost at the 1964 New York World Fair with the display of AT&T's picture telephone. Picturephone as it came to be called, was presented as a forerunner of a new telephone service that would be in every home within a decade. While the picturephone drew enthusiastic response from many in the 1960s, it was soon apparent that the cost of interactive video telephones in every household would be prohibitive for the foreseeable future. However, many began to explore commercial or institutional applications for interactive video. Thus, in the late 1960s and early 1970s, a series of pilot projects were begun using interactive video for business meetings, medical training, education and police work. These pilot projects employed a range of transmission technologies, for example microwave, cable television and special telephone lines. Also, they employed a range of communication configurations, for example two-way audio and video, one-way video and two-way audio, and data transmission accompanying audio-video communication. Many of these early projects experienced severe technical problems, as they were testing new types of equipment. Others experienced resistance from user groups who felt that the technology might replace their job. Still other applications addressed needs which did not really exist or were too costly for the institutions using them.

While the early projects experienced many problems, enthusiasm for the *possibilities* of interactive television continued to grow. One manifestation of this enthusiasm was a multi-million dollar pilot program undertaken by the National Science Foundation in the mid 1970s. The

National Science Foundation fixed upon interactive cable television as a resource for the delivery of social services and education. Their interests were tied to questions about the costs and effectiveness of delivering social services via interactive television. It was felt that if interactive television could be demonstrated to be cost-effective, it would be adopted by many communities and used to extend services to those who did not have access to them and improve services to those who received information and education in limited face to face settings or through traditional one-way media.

The National Science Foundation funded three projects, in Reading, Pennsylvania; Spartanburg, South Carolina; and Rockford, Illinois. Collectively, the three projects were intended to provide training for firemen and teachers, education for high school students, training for day care workers, and social service information delivery to senior citizens. In addition to these stated intentions, the three projects provided a range of further services. And, in each case, extensive research was conducted about the applications.

Just as the National Science Foundation projects were concluding, a commercial interactive cable project was begun with great fanfare: the Warner Amex QUBE system in Columbus, Ohio. QUBE utilized one-way audio and video into homes and simple data transmission from each home back to the cable headend studio. The principal format in QUBE programming was multiple choice responses to opinion; questions posed in programs, for example, how viewers felt about a planned shopping center in town.

While the National Science Foundation projects in Rockford and Spartanburg yielded a wealth of valuable research (Baldwin et al, 1978 and Lucas, 1978), the systems did not flourish after the research ended. Similarly, QUBE provided extraordinary publicity value for Warner Amex and helped them to win many franchises. However, use of the system (still in place in Columbus, Ohio and a few other cities) has been low and interest has generally declined (Kahan, 1983). The system in Reading, Pennsylvania stands as a curious oddity. Research about the system suggested that it provided only moderate cost-benefits for social service information delivery (Moss, 1978). Yet, it took root in the community and flourished. The reasons for this, along with some general findings about how the system has been used, are discussed in the brief case study that follows.

Interactive Cable Television in Reading, Pennsylvania: Setting Up The Organization

The Reading interactive cable television project was first and foremost a community-based social innovation. Before any equipment was purchased and before any programming was planned the project team undertook extensive organizational work within the community. This began with the formation of the consortium that received funding from the National Science Foundation. It included the local cable company, the senior citizen council in Berks County, the local housing authority and the City of Reading itself as well as the Alternate Media Center and Graduate School of Public Administration from New York University. Second, some of the NYU staff took up permanent residence in Reading during the entire course of the project; they joined the community. Then, several important organizational decisions were made to help implant the innovation within the community. A staff of technicians and camera operators who would run the equipment was hired locally. A community board was established to provide counsel and support initially and, eventually, to take over the system. Also, a programming committee of senior citizens was set up to advise on programming decisions. Finally, a nonprofit local entity was created. These organizational elements provided early credibility within the community and anticipated the eventual need for local fundraising to continue the system and a local management group to run it. In this sense, the transition process from federally funded experiment to ongoing local system was begun on the first day of the project.

System Design

Early technical design decisions contributed much to the style and conventions that evolved over time. There was a conscious effort made by the NYU project team to build a simple and modest technical configuration that would lend itself to operation by individuals with little or no technical background. It was hoped that this would place more emphasis on the quality of human interaction patterns and less on the technical demands of the system. Thus, instead of studio environments, community spaces were selected for program origination. Also, these were called "Neighborhood Communication Centers" rather than "studios." There were three Neighborhood Communication Centers in the original design. The interactions among participants at the three interconnected centers formed the basis for most programming. In addition, a few remote sites could be linked into the system when needed, for example city hall and the local social security office.

One of the early technology choices, selection of off-the-shelf black and white cameras and fast zoom lenses was directed by a concern for reducing technical problems and providing more comfort to participants. For example, the fast zoom lenses necessitated minimal adjustments to existing room lights, thus reducing heat and glare. The fabled hot lights of stage and screen are simply not suitable for ordinary people's comfort. Further, the zoom lenses allowed informal and staggered seating arrangements with no loss of image size. This became a very important issue several months into the project when a split screen convention emerged. That is, a split screen was used to show two people from different centers interacting with each other. The simple camera arrangements and zoom lenses allowed the camera operators to quickly align the image size and position on the screen of the two interactants.

In keeping with the simplicity of design, only one camera was used at each location to capture participant behavior (a separate graphics camera and videotape feed were also available for use). This camera was placed next to the television monitor that displayed the downstream video feed; that is, the image being sent out by the cable system to home viewers as well as to each neighborhood communication center. Since the monitor conveyed the video communication from other centers as well as what was "going out" to homes, participants directed their gaze towards it most of the time. And, when interacting with someone at another center, a participant spoke to the monitor, not the camera. However, since the camera was located right next to the monitor, a person speaking to the monitor gave the appearance of eye contact to a participant viewing at another center.

The design decision of one camera per location to capture participant behavior also helped everyone to understand shifts in action from one location to another. That is, any time a viewer saw a cut from one camera perspective to another, it meant that the action had switched to another center. Since three or four sites were commonly participating in a given program, the issue of "location" for a given person at a given moment was very important and potentially problematic. The one camera per location decision helped viewers identify where a shot was originating, along with occasional comments by a program host, the distinct appearance of each center and the regularity of participants at each site (that is, viewers learned that certain regular participants always went to the same center).

The video signals from each center were transmitted to the cable headend where a technician selected the shot or combination of shots (split screen) which became the downstream feed to home television sets and the video monitor at each center. The technician was called a "switcher" not a "director" (there were no directors). He was instructed to follow the action and show the person or persons who were speaking. For this reason, there were very few reaction shots or other techniques used by directors in commercial television to comment about the action. For example, in commercial television, a director might instruct a camera operator to zoom-in for a closeup shot of a person's shaking hands to show that the person is tense. This type of shot was not used in Reading's interactive programming. Further, the technician-switcher in Reading, unlike the invisible director in commerical TV, was often brought into the programming by a comment from a participant at one location. For example, a speaker commonly said, "Bruce, punch up Horizon Center," instructing Bruce, the switcher, to show a particular neighborhood communication on the video monitor.

In many ways, the configuration of the audio system was more problematic. Everyone involved felt that the ability to interrupt easily and speak informally would be crucial to successful programming. However, the presence of more than a dozen microphones spread over three or more locations, each one "open" all the time to pickup spontaneous comments, can create noise, feedback and confusion. After much trial and error, the project team selected directional, hand held microphones for use by most participants supplemented by a few lavalier microphones for use under some conditions by a program host or guest. This microphone mix helped to reduce ambient noise while allowing wide pickup in each center. The directional microphones were placed strategically throughout the seating area at each center. Participants learned quickly that they had to grab one of these microphones and speak into it in order to be heard at the other centers. The audio output at each center was enhanced by placing high fidelity speakers appropriately for each center, based on the size of the room and seating arrangements. The difficulty of finding the right balance so that participants could speak and hear without disruption to the interaction needs to be stressed. Indeed, the overall process of getting the audio to function effectively required more work than setting up the video.

Programming

Most of the programming on Berks Community Television, Reading's interactive channel, may be categorized as a talk format, with interaction taking place among participants at two to four origination sites during a given program. In the early years of operation all programming took place during the day, from 10 a.m. to noon. Programming was in black and white, and ninety percent of it was live. Individuals who wanted to participate fully in programs (be seen and heard) went to one of the neighborhood communication centers near where they lived. Those who watched the interactive programs from their home could call-in and participate in an audio-only mode. Calls from home viewers were solicited frequently during many of the programs.

Some of the early programs included: *Scrapbook*, a discussion of local history and folklore; *Inside City Hall*, an interview/discussion program between seniors and the mayor or a city council member; *Singalong*, a very popular program in which participants at all the centers sang favorite songs together; *Yoga For Health*, a five minute exercise program; *Sense and Nonsense*, a quiz program; and, *The Eban and Herb Show*, a polling-discussion show on a broad range of topics. Most programs were thirty minutes long, though a few ran a full hour and a few were five minutes in length.

A major goal of the National Science Foundation, in funding the Reading project, was to determine if interactive programming could effectively deliver social service information. This was reflected explicitly in some programs such as a question and answer session with the local director of the Social Security Administration. Other programs mixed social service information with entertainment. For example, *Sense and Nonsense*, a quiz program, mixed trivia questions, questions about Hollywood movies and social service or nutrition questions designed to educate those who might not have known the answer. Still other programming, for example *Singalong*, involved pure entertainment. A major finding of the research conducted about the Reading project was that while interactive TV was a reasonable way to convey social service information to senior citizens, an equal or greater benefit was the socialization that followed from an individual's participation in the programming. Participants made new friends, became more active, communicated more and in some instances gained renewed self esteem.

In most of the programs a strong emphasis was placed on center-to-center communication. Thus, in a discussion show with local politicians, the host of the program and the guest politician were commonly situated in separate origination sites. The seating arrangements also fostered center-to-center interaction. Everyone at a given center, host, guest and participants faced in the same direction: towards the monitor, where they would see participants from the other centers. Indeed, if two individuals who were seated side by side in the same center began a dialogue, they commonly maintained an orientation towards the monitor rather than face each other directly. The center-to-center communication flow quickened the pace of the programming, since the visual action moved from location to location in any dialogue between participants at separate origination sites.

During the first few months of system operation, programs did not keep to a rigid schedule. A thirty minute program might run from twenty-eight to thirty-two minutes. Along with this relaxed sense of time, programs often began and ended with a few minutes of chitchat totally unrelated to the topic or purpose of the program. Within a few months, scheduling requirements of the cable system (the interactive programming was bracketed by commercial programs that began and ended within precise time constraints) and a general concern about managing the interactive programs led the project staff to adopt more precise time schedules. However, the informal chitchat feature (which almost everyone liked) was maintained in a more controlled way. Indeed, it evolved into a convention of "sweeping" the centers. At the beginning and end of the program day as well as at the beginning of many individual programs the host asked the technician-switcher at the cable headend to show each center one by one. During this sweep of the centers, participants frequently smiled or waved, greeted a person at another center, asked about someone's health, etc.

The issue of appropriate structure for the interactive programs was from the earliest days a major issue for all. A number of competing models existed. The programs could attempt to imitate commercial TV; treat the participants' interactions like other community social gatherings such as church or town meetings; or, explore totally new formats that built upon the unique characteristics of the interactive technology. By and large, the NYU team wanted to explore new interactive formats; most seniors recognized that they were involved with a new technology but also fell back naturally on formats and behavior borrowed from their social experiences with town meetings, church, etc.; and a few seniors envisioned themselves to be a Johnny Carson or Phil Donahue.

These competing models were readily apparent in early programming, where there was much variation in format and style. Within a short time however, many program conventions emerged based upon the characteristics of the technology and system configuration, for example, the center-to-center communication flow. Competing formats based upon the seniors' social experiences, for example town meetings, were somewhat less influential as the new interactive conventions emerged. It may also be argued that the new technology and existing social formats merged, as in the interview/discussion program between seniors at three neighborhood communication centers and the mayor at city hall. The influence of commercial TV is more complex. Those who tried to imitate commercial TV formats by reading from a script fared poorly and the practice was soon discouraged. Other commercial TV practices such as opening a program with a standard video sequence led to parallel practices: displaying a designed graphics card and playing a short musical theme at the opening of some programs. And, the aspirations of a few to imitate the style of Johnny Carson or Phil Donahue remained. Indeed, while Berks Community Telelvision has evolved into a clear and distinct form of television, with its own set of formats and conventions, the subtle influence of commercial TV models in the minds of producers and participants will likely always remain.

Communicating on Interactive Television

How do ordinary people, in this case senior citizens, adapt to a new communication technology like interactive cable television? Berks Community TV provided a fertile ground in which to examine this question. In Reading, people came to the new situation with existing communication patterns based upon their experiences in everyday face to face situations and with other technologies such as the telephone. Early usage of interactive television in Reading demonstrated that people initially borrow behaviors from other, more familiar situations and treat the new communication form as if it were the more familiar form. Thus, some individuals after completing a speaking turn would conclude with "over," a pattern borrowed from amateur radio operators. One participant, after speaking into a microphone, placed the mike next to his ear in order to hear the response. He treated the microphone as if it were a telephone receiver. This pattern is not new or unique to Reading. Alexander Graham Bell is reported to have begun his telephone conversations by saying "Ahoy."

Since individuals in Reading borrowed from different areas of experience, there was much variation in the early communication patterns on Berks Community TV. Over time, as people learned about the system and shared codes of behavior evolved, communication patterns became more consistent and efficient. They were more consistent in that most participants said or did similar things in order to communicate a given meaning. They were more efficient in that they could communicate a given meaning in less time and did not require the same degree of redundancy to make sure others understood the message.

Interactive programming in Reading also demonstrated how many of the fundamental prerequisites for a communication exchange can become disrupted when people exchanging messages are separated physically. For example, in a face to face situation it is relatively easy to establish that someone else is in your presence and therefore able to communicate with you; whether they are talking to you or someone else; who they are; and what you are talking about. With these prerequisites in place, communication can flow relatively efficiently between individuals. On interactive TV, individuals have to do more work to establish these prerequisites. Moreover, the prerequisites once established are subject to disruption, for example, a technical problem at one of the neighborhood communication centers can eliminate someone from your "presence." Similarly, a comment seemingly addressed to everyone at all the centers might be a response to a side comment spoken off-mike and heard only by those at one center. These communication issues are not unlike those we experience when using a telephone. However, an elaborate and shared set of conventions has evolved for managing telephone interactions. Many of these conventions are unexamined because they seem natural to us at this point. For example, the occasional "uhuh" of a listener in a telephone conversation not only signals 'I agree' or 'I'm following what you're saying', but also 'I'm still on the line and in your presence.' If a listener in a telephone conversation does not provide these minimal signals of continued presence on the line, a speaker is likely to inquire after two or three minutes, "Are you still there?"

On Berks Community TV, much of the work to establish and maintain the prerequisites for communication has evolved over time as a job for the program host and a secondary host at each center. They help to establish who is speaking, what the topic is, when someone leaves the room, etc. In addition, the television monitor in each center serves to establish and maintain the communication channel between interactants. It conveys who a person is interacting with; where the other person is; and what information is being transmitted throughout the system. For all these reasons, the monitor has served as a powerful regulator of interaction.

The communication behavior of participants on Berks Community TV was also affected by the regularity of the interaction, the narrow geographical locale of participants and viewers, and the synchronous relationship between programming and day to day life. That is, the programming was live and took place each weekday. In addition, each program had many regular participants; participants and viewers all lived within a few miles of each other; and nearly everyone shared knowledge about local streets, stores, schools and people. Thus, if a regular participant had a cold *yesterday* during a program, another person could ask during a program on the following day if he were feeling better. No such assumptions can be made about programming which is prerecorded and broadcast or cablecast over a variable time period.

Communication patterns in Berks Community TV were also characterized by the use of specifics rather than generics("You can get it at Boscov's" rather than "You can get it at your local drug store"). Participants assumed that the audience shared their knowledge about Reading. This contrasts markedly with most programming on commercial television where accepted communication conventions eliminate most temporal references (news is an exception), frequently employ generic references rather than specific ones, and assume little shared knowledge by the audience (major public items of knowledge such as the name of the U.S. president or a football star are an exception). This also means that much of the programming on Berks Community TV would be difficult to follow for a cable audience in Houston or Boston.

Evolution of The System

To an observer familiar with the early Berks Community Television programming, there has been relatively little apparent change. In 1977, evening programs were added to the schedule. They emphasized community wide issues and included many participants in addition to senior citizens. In 1981, Berks Community TV changed to color and added a new "studio" space in downtown Reading. Over the years the number of origination sites has expanded to more than a dozen. Additional high schools have linked into the system as well as local colleges, hospitals and more government agencies.

In 1977 national funding for the Reading "experiment" ended and local funding as well as local management took over the system. Since then, eight years of community support have strengthened the position of Berks Community TV within the City of Reading and surrounding areas. The current annual budget of $250,000 comes from a healthy mix of sources, including the City of Reading, the County Office of Aging, local corporate and foundation grants, the local cable company and viewer donations. Most of the budget goes for seven full time and two part time staff. The paid staff is supplemented by an army of local volunteers, both technical and programming.

The programming conventions appear to be well embedded. While an interview style borrowed from commercial television is somewhat more evident, the continued use of split screens and a center to center information flow maintain an overall style that is quite similar to programming in the early system. Many of the original programs and program producers persist. Their expertise and participation in the now expanded system has helped to provide a continuity of style for the new programs and user groups. For example, Oda Miller, a retired social service worker who joined the project at its earliest stages has continued to contribute programming concepts, interviewing and hosting responsibilities. So too has Gene Shirk, ex-mayor of Reading, chairman of the Berks Community TV board and host of *Generation Gap* from its beginning. It was Gene Shirk who developed the idea of opening the system to young people and in this way, began the transition to a broadly-based community channel. Many other early participants have continued as program producers, hosts and regular participants. Mae Fieck, who hosted the original interview program with city hall politicians; Ed Yost, who hosted *Scrapbook*; and Marie Pelter, who hosted the quiz show *Sense and Nonsense* have all since died. When Marie Pelter died, her obituary in the local newspaper ran under the headline "TV Personality Dies." This was one of many indicators that those who created and maintained Berks Community TV have achieved both respect and notoriety within the community.

Conclusion

Having outlined a brief case history of interactive television in Reading, Pennsylvania, it is appropriate to return to the questions we posed earlier: why has the Reading system flourished while many other interactive systems did not and why has Reading not been replicated elsewhere? The two questions may be interlinked.

The apparent success of Berks Community TV has relatively little to do with interactive technology as such. Berks Community TV flourished, first, because the timing and location for the service were propitious. The conditions of time and place which fostered growth and acceptance for the service were a heavy cable penetration, no local broadcast television and a cohesive community of viewers. Reading, Pennsylvania is one of the "gritty" cities in the Northeast whose population has dwindled but which holds deep traditions of railroading, textile mills and Pennsylvania Dutch ancestry. Served by television in Philadelphia and Harrisburg, the community naturally yearned for "local" programming. Also, local government agencies with a mandate to provide information services to the public and funding to pay for them found Berks Community TV a good way to do the job. It did not hurt, as well, when a local city councilperson who appeared regularly on early programs and won popularity through that participation, was then elected mayor of Reading at the next general election.

Second, while the lure of advanced technology helped to secure early funding from the National Science Foundation, the implementation of the system was not technology driven. The needs and wants of people participating in the programs, watching as part of the home audience, and serving on the many committees associated with the system were emphasized above all technical considerations. This helped to create a loyalty and tenacity that was crucial to the ultimate survival of the system.

Third, the continuation of the system has been tied to the skills of those who have run it for the past several years, a manageable budget, and sources of local funding. Further, there is no guarantee that Berks Community TV will continue for another decade. Like other community services, it is subject to changes in the local environment as well as the resources and desires of people who provide, consume and pay for the service.

Berks Community TV has not been replicated elsewhere, nor should it, because its system design, program style, user groups and content have been specifically tailored to conditions in Reading. If there is a lesson in Berks Community TV for other communities, it is to begin with a clear understanding of local conditions, needs and resources, then to build communication services based upon that understanding.

Finally, Berks Community TV may be used to discuss some fundamental questions about interactive television. Specifically, is interactive television thrown into the mix, just a new format for normal one-way television, or a new medium entirely? This is not simply an academic issue. If we treat interactive television as a form of face to face interaction, users are likely to place demands on the system to do what people can do in face to face situations. No telecommunication system can duplicate an exchange that is mediated only by the human sensory apparatus. Interactive television will, under these circumstances, fall short of the standard we match it against. Conversely, if we adopt one-way television as a general model for interactive TV, users will likely imitate program formats and styles of behavior present in commercial television. This presents at least two obstacles. First, few interactive TV systems could match the production budgets of commercial television. Programs that attempt to copy commercial TV formats may therefore look like cheap imitations. Second, the likely users of interactive television are not actors and will not be able to match the performance of commercial TV actors.

An alternative approcach is to carefully examine the characteristics of interactive television systems and to explore the kinds of communication for which they are best suited. Berks Community TV provides some examples of suitable applications but it by no means exhausts the possibilities. There are many examples of successful interactive television applications other than those in Reading. Collectively, the work of those who create interactive programming and the behavior of audiences that participate in it will forge a set of shared conventions. In this sense, it may be argued that interactive television is a distinct communication medium that is in the process of being formed.

References

Thomas Baldwin, Bradley Greenberg, Martin Block and Nicky Stoyanoff, "Rockford Ill.: Cognitive and Affective Outcomes." *Journal of Communication*, Volume 28, Number 2, Spring 1978, pp 180-194.

Hazel Kahan, "The Cable Subscriber Speaks: Channels, Choice or Chaos?" New York: Advertising Research Foundation, December 1983.

William Lucas, "Spartanburg, S.C.: Testing The Effectiveness of Video, Voice and Data Feedback." *Journal of Communication*, Volume 28, Number 2, Spring 1978, pp 168-179.

Mitchell Moss (ed.), *Two-Way Cable Television: An Evaluation of Community Uses in Reading, Pennsylvania.* New York: The NYU-Reading Consortium, 1978.

SLOW DOWN
IMAGES MAKING
TIME MONEY
IDEAS COLD
TUNE MEDIA
FICTION THINK
NEED CONSUME

WHAT IS VIDEOTEX?

Vincent Mosco

The Fundamentals: Teletext

TELETEXT AND VIDEOTEX are the current generic terms to describe data transmission systems that bring information to viewers over a modified television receiver. Data is typically transmitted over broadcast television channels (teletext) or phone/cable circuits (videotex) (Figure 1).

Teletext is the term commonly applied to the broadcast version of videotex. It is essentially a one-way system that supplies digital data on the normal broadcast signal by placing messages in the unused lines of the standard television signal, also known as the Vertical Blanking Interval (VBI). The VBI is the dark bar that sometimes appears when a standard broadcast television picture "rolls," requiring the viewer to adjust a dial. A viewer with a television modified for teletext has a control keypad. This comes in a variety of forms, though the typical keypad resembles a hand calculator. The teletext viewer pushes keypad buttons to turn off regular television fare and display frames of text or picture material. One typically starts with an index that indicates what is available on different "pages." Figure 2 is a replica of one such frame taken from the British Ceefax teletext service broadcast over BBC Channel One. Selecting page 101 on the keypad brings the viewer news headlines. From there, he or she can access the details of specific stories or move to another story. Teletext is considered a one-way system because the viewer is restricted to information that the station decides to transmit. While such information is regularly updated, with updates based in part on viewer selections, there is no way for the viewer to request information in addition to or different from what the station offers.

To understand the basic workings of a teletext system, let us begin with the information—the News Headlines appearing on page 101 of Ceefax BBC-1. The information is edited in a studio and coded in digital form for transmission at a rate that an ordinary TV set can handle. The transmission or bit rate is important because a low rate means a longer turnover time for pages, and thereby causes the viewer to wait longer for information. European systems can use higher bit rates for teletext than can North

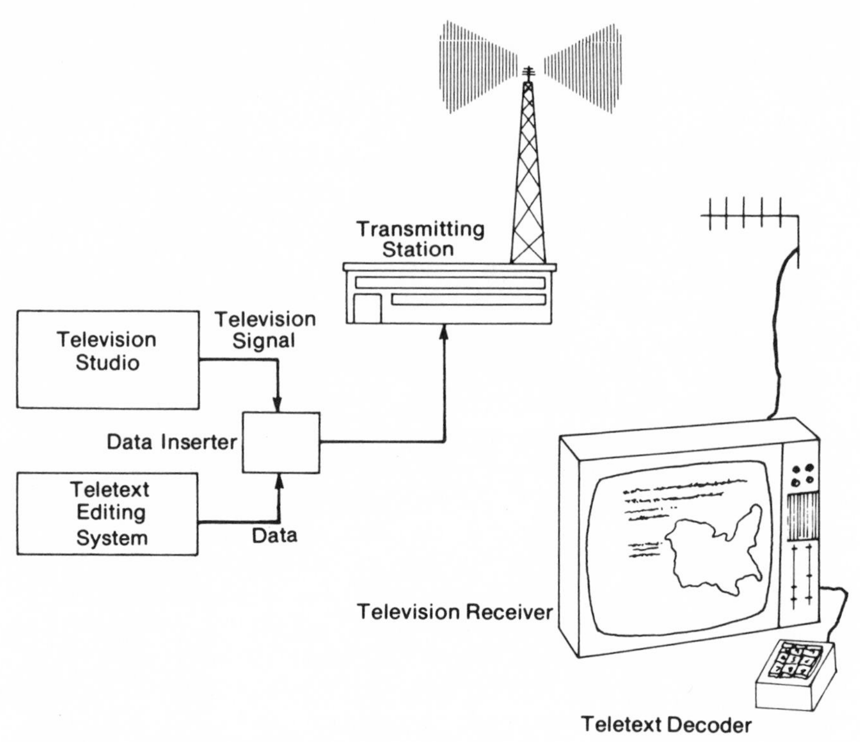

FIGURE 1. A broadcast teletext system.

American systems, because European television channels are "wider," i.e., they use more bandwidth. The information, now encoded in digital form, is inserted into the television signal on unused lines or the VBI. The more lines of VBI, the more information and the faster the viewer access time. Some teletext services use a full channel, set aside solely for teletext use. Cable television systems with 32 + channels more commonly make use of full-channel teletext than do standard broadcast systems with under a dozen channels. The digital signal can be detected by a decoder built into the television receiver or wired as an accessory to an ordinary television. The decoder receives, stores, and ultimately displays the data on viewer command. That command is signalled on the control keypad which directs the decoder to transfer the desired page to the screen. A character and graphics generator built into the decoder transfers the page to the screen. The page is removed when updated information, such as an hourly News Headline edition, is provided.

The teletext "newspaper" or "magazine" of pages is transmitted continuously. The recycle time of each page determines how long a viewer has to wait before receiving the News Headlines, weather report, or recipe that his or her keypad selection called for. If the magazine contains

200 pages and a new one is transmitted every eighth of a second, then the maximum waiting time for a particular page is 25 seconds. Access time on the BBC-1 and BBC-2 services is about 12 seconds for 250 pages of material and about 25 seconds for the 375 pages transmitted over the commercial channel.[1] Most people are accustomed to access times considerably under these figures—the time it takes to turn the page of a book or flip through a newspaper, for example. This time is an especially important parameter that teletext operators would like to minimize. Several factors in addition to bit rate and page volume determine access time. These factors include the number of television lines dedicated to teletext reception. Here, as in channel bandwidth, Europeans have the advantage. European television receivers contain 625 lines, 100 more than their North American counterpart. The less adequate 525-line system is the price that North Americans have to pay for permitting RCA and CBS to pressure the FCC into the early acceptance of broadcast television standards. European receivers are better equipped to receive teletext information in shorter access time.[2] A final major factor shaping access time is the amount of memory in the decoder. The accelerating decline in the cost of computer memory, due to advances in solid state electronics, will allow for faster access to stored information.

Since teletext relies on existing television sets, on equipment that almost everyone has in the home, teletext promises to develop into a mass medium. But this does not mean that everyone equipped to receive teletext will have access to the same information. For inexpensive variations in transmission code could limit access to certain pages to users with decoders specifically designed to receive those pages. This opens the way for scaled pricing of teletext services, a more refined version of the add-on characteristic of current cable/pay-television franchises. In Arlington, Virginia, I could buy a basic cable service for $10 a month. If I want Home Box Office, a noncommercial moviechannel, or a "culture" channel featuring opera, ballet, symphony music, I would pay another $10 for each. Certainly, one can foresee teletext "Closed User Groups" for special needs users such as travel agents, doctors, etc. But it is also likely that information useful to a wide segment of the population, on insurance companies with the best performance ratings or on savings accounts that pay the highest interest, will only be available to those with the money to pay for the information. We return to this important social consequence of teletext below. At this point it is important to understand that teletext technology holds the same potential for intensifying (or rectifying) the maldistribution of information as does the more technically advanced videotex service.

FIGURE 2. BBC-1 ceefax program index.

Contents: Page 100 (BBC-1)

NEWS

Headlines	101
News in detail	102–116
News Diary	117
People in the News	118
Charivari—a lighter look at the News	119

FINANCE

Index	120
News and Reports	121–126
Market Reports	127–129
FT Index	130
Stocks and Shares	131–133
Exchange Rates	134–136
Commodities	137–138

SPORT

Headlines	140
Sports News	141–159

CEEFAX provides a rapid service of news, results and background. CEEFAX Sports Specials, covering major sporting events, begin on 151.

FOOD GUIDE

Headlines/Index	161
Shopping Basket	162
Meat Prices	163
Fish Prices	164
Vegetable Prices	165
Fruit Prices	166
Recipe	167
Farm News	168–169

ENTERTAINMENT

Today's TV—	
BBC-1	171
BBC-2	172
ITV	173
Radio highlights	174
Films on TV	175
Top Twenty	176
Theatre	177
Opera/Ballet	178
Viewers' Questions	179

WEATHER AND TRAVEL

Headlines/Index	180
Weather Maps	181
Temperatures	182
Temperatures	183
Travel News	184–189

NEWSFLASH 150

Turn to this page to watch television program—when something important happens a NEWSFLASH will appear on the picture.

ALARM CLOCK PAGE 160

This page can change every minute. It can also be used as a silent alarm clock.
Turn to page 160 for instructions.

SUBTITLES 170

The BBC is experimenting with various ways of subtitling programs. This page shows how subtitles could look.

LATEST PAGES 190

As each new page is put in the magazine it is also put on 190 where it alternates with a news summary.

OTHER PAGES

News about CEEFAX	191
Engineering tests	197–198

FULL INDEX

A–F	193
G–O	194
P–Z	195

WANT TO KNOW MORE?

Write to:
CEEFAX Newsroom (7059)
BBC Television Centre,
LONDON, W12 7RJ

Source: Colin McIntyre, "Teletext in Britain: The CEEFAX Story." In Efrem Sigel (Ed.), *Videotext: The Coming Revolution in Home/Office Information Retrieval.* White Plains, NY: Knowledge Industry Publications, 1980, p. 36.

The Fundamentals: Videotex

Videotex, or Viewdata as it is less frequently known, allows a subscriber interactive use of large amounts of information. Rather than simply *selecting* from the 100 or so pages that the teletext service broadcasts, the videotex user can process thousands of pages of data, contribute to an existing data base, and conceivably communicate this action to other subscribers, thereby multiplying the connections exponentially. More concretely, videotex supporters see the service opening the way to new forms of learning, shopping, banking, communicating, and indeed, working.[3] The basic components of a videotex service include:

1. Information retrieval and display terminals.

These can be ordinary color television sets instrumented with a decoder to translate digital signals into a visual display. They can also be modified computer terminals equipped with a color display. The retrieval-response device can range from the teletext calculator-like keypad to a full typewriter-like unit that permits vastly more commands than the 10-digit keypad allows.

2. Transmission lines for interactive communication.

The distribution mechanism for a videotex system can be the phone lines that currently comprise the public-switched telephone network, a cable television coaxial cable network, a communications satellite system, microwave facilities, and combinations of these. The greater reliance on wired distribution makes possible the circulation of many times more information than is possible with a broadcast teletext service.

3. Computer hardware and software.

The TV or related hardware display device is linked to a computer with a storage capacity of thousands of pages. Specific programs permit users rapid access to this information, as well as record the activity of users for billing.

In a number of respects, the development of videotex is a far riskier proposition than teletext. It is by no means certain that the costs in computer hardware, programs and production, editing and regular updating of information can be offset by user payments. Videotex is not as obviously a mass medium as is teletext. Will advertisers support a service whose principal values appear to be the ability to pursue individual information interests and to interact with other subscribers? Of course, videotex offers more opportunities for Closed User Groups than does teletext. It is quite possible that multinational corporations, from banks to tourist companies, will find attractive the rapid, flexible, and private networks that videotex allows them to construct. The result would be a scale of users ranging from basic teletex services, *Time* magazine on the air, with mass appeal and mass advertiser support, to high-priced, specialized videotex

FIGURE 3. A videotex system.

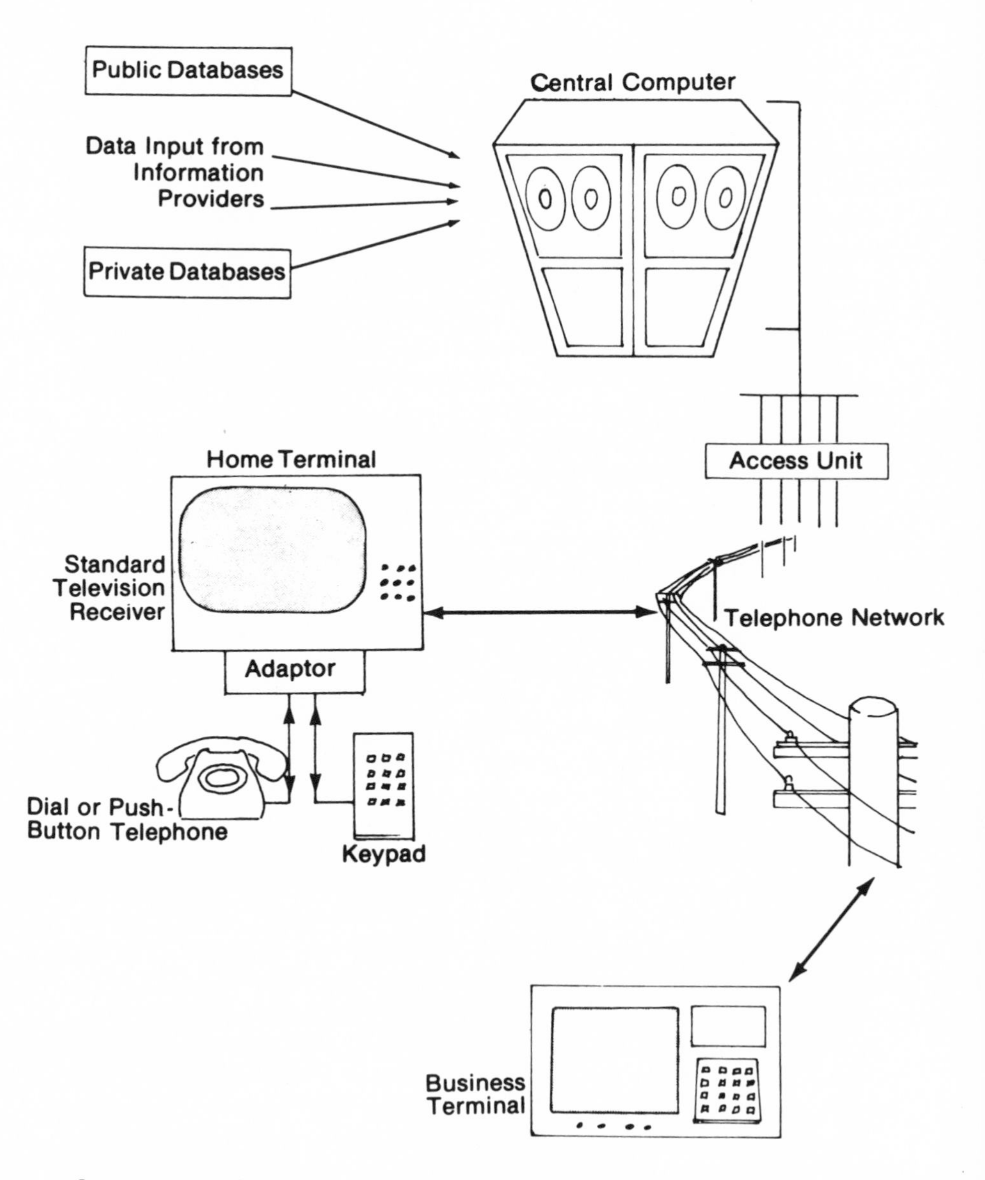

Source: Institute for the Future, *Newsletter*, 1979, 2 (1), p. 2.

services with appeal to large corporate users. The distribution of information would closely follow this scale, so that those with the money to support the more expansive videotex systems would have access to far greater communications/information resources than those without. The gap between the information rich and poor would grow.

Development of Teletext/Videotex Systems

With all due respect for the arbitrariness of technological beginnings, it is useful to start a discussion of the development of videotex in both its broadcast and wired forms with the British Ceefax experience. The British have led in the research and commercial application of videotex, though developments in France, Canada, and Japan have not lagged far behind. U.S. companies have been slow until recently to move on this specific product of the information age. Though as we see below, these companies are now taking advantage of years of experience with the building blocks of videotex to narrow the gap. Let us explore in more detail the development of videotex services and current applications in these and other leading videotex countries.

Teletext services in Britain grew out of the interest that BBC engineers had in providing broadcast services to the deaf and to Britain's several language minorities. This interest led to the development of "closed captioning" which would provide subtitles to people with decoding devices. By 1971 this specific application grew into proposals for a more general service based on the broadcast teletext principle. Advances in computer technology and data transmission led a BBC research group to come up with the basic elements of a teletext system by 1972. At about this time, a team working at the British commercial television station, the Independent Broadcasting Authority, developed a system similar to the BBC's newly dubbed Ceefax (for "see facts" or "BBC-facts," depending on whom you ask). Not to be outdone in the creative naming department, IBA called its teletext service, ORACLE, short for Optional Reception of Announcements by Coded Line Electronics.

The first CEEFAX lab trials took to the air in 1973. Satisfied with the technical success of these tests, BBC management gave the okay for a two-year market trial to begin in September 1974. This trial also convinced enough of management that teletext contained substantial market potential to get government approval for a commercial system to begin toward the end of 1976.

There are currently three telecast chanels in use throughout Britain. The BBC operates two, and the third is under the management of the commercial IBA. Teletext experienced its sharpest growth in Britain beginning in 1980. From 40,000 teletext sets in use at the start of that year, the population grew to 150,000 by March 1981. Since each set is estimated to be used by three people, teletext "readership" reached the half-million point by mid-1981. While representing substantial growth, these figures are under the estimate of the British Radio Equipment Manufacturers Association which had anticipated 300,000 sets and 900,000 users by the end of 1980.[4]

The teletext receiver costs about 120 pounds more than a standard television set. Decoder costs are declining thanks to the mass production of semiconductor chips. Nevertheless, the majority of teletext sets are rented in Britain, a pattern that is common in standard television as well. This pattern makes Britain particularly well suited to the growth of video innovations. Most people view teletext not as a 120-pound additional initial investment, but as a one-pound additional weekly charge.[5] Despite the popularity of teletext rentals, a detailed study of teletext users reveals that:

they come from the professional or skilled occupational classes, are generally more prosperous than the population at large, and are generally heavier users of information of all sorts, including libraries.[6]

The BBC has 20 people running both its teletext services on an annual budget of 200,000 pounds. These funds, as well as the 500,000 pounds in capital investment, have come from the national license fee that supports the entire BBC. Standard television viewers have therefore subsidized the development of the BBC teletext service. A 1981 Act of Parliament opens the way to teletext advertising. Under the Act, the IBA is permitted to sell 15% of its pages to advertisers.

The mainstays of UK teletext content are the wire service news, sports, weather, travel information, TV/radio program listings, light entertainment, and captioning for the deaf. Market research studies indicate that the average teletext subscriber makes use of the service for about two hours a week.[7]

Prestel

After five years of development, British Telecom (once the British Post Office Telecommunications Division) launched the first commerical videotex service in September 1979. Prestel reached 13,000 sets in use by October 1981 for a total readership of about 25,000. This makes Prestel the world's largest computer-based information service. Nevertheless, growth is not running at anticipated exponential rates. Rather, a slow though steady increase of 500-600 sets a month was reached by mid-1981.

Unlike the Ceefax service, most Prestel subscribers are businesses. As of October 1981, 87% of all users fell into the business category. Twenty percent of all Prestel subscribers are travel agents. This represents 40% of all travel agents in Britain. The other major subscribers are in the fields of investment (*The Financial Times* was an early information provider), computers and electronics, print publishing, and education.[8]

The average use of Prestel is for about nine minutes a day or 142,000 frames of information, up from a 1980 average of 56,000 frames. Excluding local telephone charges, the average Prestel user pays about 100 pounds a year for the service. The price of a Prestel set ranges from 200 pounds for a system with a simple numeric keypad to 1,500 pounds for one with a typewriter-like alphanumeric keyboard. Sets rent for about 25 pounds a month.

By October 1981 some 650 British and foreign Information Providers (IP) had booked 200,000 Prestel pages and had actually filled 180,000 of them. The most successful IP earns about 50,000 pounds a year. Prestel collects this revenue and pays it to the IP after deducting a 5% service charge. Not all IPs are earning revenues. According to an executive from one of the more successful IPs, success is only achievable through a tenacious market orientation.

We are now in the real world, and the fantasies of the test service era have faded. The information provider community is polarizing, leaving in the business those who are taking a proper and commercial approach with professional application and who can invest properly in a new medium. It is a pity, perhaps, that there is no cheap and universal channel for the small and the worthy, but computers and data storage cost money, the telephone network does not operate itself, the common indexing and other central services need people and editing terminals cannot be obtained on charity. There is a price to be paid for everything.[9]

"There is a price to be paid for everything." If there is one sentence that captures the views of people participating in the first commercial videotex operation, it is that. There are enormous pressures on Prestel producers to reach a threshhold of market penetration that will establish the viability of videotex services. These pressures are particularly intense because the British government appears to be prepared to let Prestel sink or swim on its own commercial ability. Such pressures filter down to the producers of videotex information. Information providers succeed or fail on their ability to convince enough people to pay the frame fee that Prestel clearly indicates on each page of information. As a result, the system is skewed in the direction of marketing services to those who can afford to buy both sets and information. Prestel therefore means updated commodity market prices and not the location of social service agencies and health clinics.

It is too early to offer a complete assessment of Prestel. However, the direction of development is clear: The need for immediate commercial success means cutting back on earlier mass marketing efforts and an emphasis on specific business uses.

France: Videotex as the Cornerstone of Telematique

The French approach to the development of telecommunications technologies differs from the British in that there is more explicit government direction to the French effort. In fact, videotex represents one part of a wider government program in *telematique*, a neologism coined by Simon Nora and Alain Minc in their influential report to the President of France on the *Computerization of Society*.[10] This report laid the groundwork for a substantial government investment in the French telecommunications industry. Nevertheless, while the degree of government planning and involvement distinguishes the French from the British effort, the purposes are similar. The most important condition that the Nora/Minc report calls on the French government to meet is:

First, the increase in competitiveness, matched with an industrial policy adapted to the new international division of labor, must generate an increase in markets. The restoration of foreign financing will make it possible to stimulate new domestic demand, thus increasing employment. Financing could then come from the excess in productivity.[11]

One basic element of the French telematics program is modernization of the nation's long-derided phone system. From 1975 to 1980 the French government organized a $30 billion program that has increased the number of lines from 7 to 16 million. The government plans to continue the sharp growth in equipment and further modernize the system with the introduction of digital exchanges. The French hope to maintain the 5-7% increase in equipment and thereby match the anticipated West German phone population of 24 million by 1985, and 30 million by 1990. Furthermore, current plans call for adopting digital switching in half the system by 1985 and over three fourths by 1990.[12] This will provide a base for enabling subscribers to take advantage of the telematics products that French industry is producing. Ironically, the French are now benefitting from their earlier lack of telecommunications development. With few products and vested interests to overturn, the way is clear for France to establish an advanced mass network.

The first major advance into that telematics field is the introduction of mass-market products including videotex, teletext, remote copiers, and point-of-sale terminals. The element of this program that has received the widest attention is the government plan to offer free to every phone subscriber in France an interactive videotex terminal whose principal initial use will be to provide a "Yellow Pages" Directory Service. This may make France the first country in which most homes will be equipped with a computer terminal. The current plan is to provide 30 million terminals by 1992, though industry estimates anticipate more on the order of 8-10 million by that time.[13] Such terminals could be adapted to offer teletext, videotex, and other related services.

Independent of the Electronic Directory, the French have put together an ambitious program in teletext and videotex. These two services are more closely connected in the French experience than they are in Britain. This is in part the result of a joint research effort sponsored by TDF, the national television network, and the PTT, which operates the French telephone service. The result is a teletext/videotex system that is fully compatible. The ANTIOPE teletext service has operated on a small scale since 1977 in Paris and Lyon. It offers a magazine format similar to the CEEFAX service. ANTIOPE has been marketed aggressively outside of France. CBS was an early supporter of the system, to the extent of petitioning the FCC to make ANTIOPE the American standard for teletext.[14] In October 1981 the French government introduced Inteltext, an enhanced teletext service that will offer both general and special-interest information in a U.S. market trial.[15]

TELETEL, the French videotex service, is currently undergoing its first major field trial in 2,500 households in Velizy, on the outskirts of Paris. The mix of IPs for the Velizy trial includes the expected concentration of businesses, from travel agencies to mail order firms and investment services. Though in keeping with the prominent role that the Catholic Church has played in French history, the Church is offering the first Pray by Videotex service.[16] Three hundred of these 2,500 households are being provided with "smart cards" or microchips on a card. These will enable families to pay for advertised products and services in the same way that "money-cards" are used in automated teller devices for electronic banking. The only difference is that in the Velizy test the card user will not have to leave home to complete the purchase. Experiments similar to this involve the application of videotex to the automated office, factory, and, in a few cases, to work at home or "teleworking."

The Electronic Directory and other videotex services that are consumer-purchase oriented give a definite direction to the French telematics program. The French are essentially involved in a massive effort to build a state-of-the-art infrastructure for the electronic mass consumption society. This is the basis of both home and business applications. From teleshopping to teleworking, the market connection is clear. Nevertheless, the entire program is currently under review by the Mitterand government. Initial responses indicate possible moderation in the pace of the program, but not in its direction.[17]

Canadian Telidon

The Canadian government's involvement in telecommunications is not as ambitious as the French program in telematics. Nevertheless, Canada has followed the French pattern with a substantial government commitment to the growth of a national videotex service that might compete successfully for lucrative foreign markets (spelled U-N-I-T-E-D S-T-A-T-E-S).

The Canadian Department of Communications developed an R&D program that began to produce a videotex product technically superior to Prestel in 1978. This resulted in Telidon, a service that may not be a full generation ahead of existing systems but represents an advance in database flexibility and graphic-display capacity. Telidon has enjoyed a substantial Canadian government financial commitment. In February 1981 the Department announced a two-year $27.5 million outlay for Telidon development. For a country with one tenth the population of the United States, this investment, on top of the $12.5 million that the Department has already committed to Telidon, represents a major national effort. The fundamental purpose behind this effort differs little from that of the British and French. According to Communications Minister Francis Fox, this spending "will be to ensure the existence of a commercially viable videotex industry in Canada with a capability to compete in export markets."[18]

The ambitious nature of the Canadian effort to develop videotex markets is suggested in Table 4, a description of Telidon field trials. These trials cover most of Canada's provinces, as well as locations in the United States and Venezuela. They are sponsored by both private and governmental agencies, cover rural, town, and urban areas, and offer data bases in both official languages of Canada. Morever, there is diverse distribution of terminals (homes, businesses, public places, and schools) and several different transmission modes including satellite, TV broadcast (over VBI), cable, paired wire, fiber optics, and ordinary telephone. A number of developments in 1981 gave Telidon the possibility of developing that preferred position in the international market. In February 1981, both Time, Inc. and the Times-Mirror Company announced the selection of Telidon for videotex projects. Time, Inc. is using Telidon to test the first national multiple channel teletext service in the United States. Times-Mirror a media conglomerate that includes the *Los Angeles Times* among its many holdings, is using Telidon for its L.A. videotex trials. Finally, in May, AT&T announced standards closely compatible with Telidon for proposed videotex services.

The Canadian program is unique in the extent of public involvement and social application in these field trials. The Department of Communications guarantees public and educational representation on its videotex

TABLE 4. Telidon systems, current and planned.

Project Name	*Location and Start-up*	*Number of Terminals and Transmission Method*
TV ONTARIO	Throughout Ontario January 1980	55 Telidon Terminals Broadcast teletext via TV Ontario and Anik B satellite Videotex via telephone
PROJECT IDA	South Headingly, Manitoba June 1980	33 Telidon Terminals Videotex via coaxial cable
PROJECT VISTA	Toronto, Montreal, Quebec City May 1981	491 Telidon Terminals Videotex via telephone
PROJECT MERCURY	Saint John, N.B. April 1981	45 Telidon Terminals Videotex via telephone
VIDEOTEX PROJECT	Vancouver, B.C. Summer 1981	150 Telidon Terminals Videotex via telephone
PROJECT ELIE	Elie, Manitoba Fall 1981	150 Telidon Terminals Videotex via optical fiber
TELIDON II	Montreal, Quebec Fall 1981	250 Telidon Terminals Teletext and Videotex by cable, Telidon II data base accessible by keyword search
PROJECT GRASSROOTS	Southern Manitoba April 1981	25 Telidon Terminals Videotex via telephone
TASK FORCE ON SERVICE TO THE PUBLIC	Across Canada May 1981	20 Telidon Terminals (eventually 100) Videotex via telephone
CBC TELIDON	Toronto and Montreal 1981	TBA Broadcast teletext
CABLECOM	Regina and Saskatoon, Saskatchewan, 1981	TBA Videotex via telephone
AGT TELIDON	Calgary, Alberta July 1981	30 Telidon Terminals Videotex via telephone
MARITIME TEL AND TEL	Nova Scotia Early 1982	TBA
TELEGLOBE TELIDON	International Summer 1981	50 Telidon Terminals Videotex via telephone and data switch network
WETA/AMC TELETEXT	Washington, D.C. June 1981	60 Telidon Terminals Broadcast teletext via PBS station WETA
VENEZUELA PROJECT	Caracas, Venezuela April 1981	26 Telidon Terminals Videotex via telephone
TIMES-MIRROR	Los Angeles, CA Late 1981	200 Telidon Terminals Videotex via telephone
TIME-LIFE	TBA	TBA

Source: Government of Canada, Department of Communications, *Telidon Reports*, July 1981, (6) pp. 5–7.

bodies. The Department has funded more than the standard market-viability studies. It has supported among the few analyses of public access and social impacts.[19] Moreover, schools are more closely involved in Canadian field trials than in either Britain or France. TV Ontario, the provincial educational network, is particularly active in development of a provincewide education service. Furthermore, Canadian government agencies, notably Statistics Canada, are involved in expanding ways to ease public access to government data through the location of videotex terminals in such public places as post offices and shopping centers.

Several explanations for these unique sociopolitical features come to mind. One is that Telidon is at the field trial stage. The largest trial in Table 4 involves only 491 terminals. Recall that Prestel has over 13,000 terminals in full commercial use. Neither the Canadian government nor Canadian business have had to confront the "price to be paid for everything" reality that Prestel participants make painfully explicit. Moreover, such public participation in the early stages of Canadian videotex is itself good business. Just as the "Golden Age of TV" was a successful industry strategy to sell sets, the promotional subsidies of the Canadian government may well reflect the sober business judgment of people aware of what it takes to create a market.

The United States: Leaving it to the Market

In the U.S., videotex grows out of the hardware, software, and interests of the telecommunications *powers that be*. The television screen on which CBS has built a media empire is the basic videotex display device that CBS is using in its field trials of the service. The television is linked to computers that have made IBM a global power. The linking device is the telephone line—the conduit through which AT&T has created the largest business corporation in the history of the world. As we shall see, there are variations on this television-computer-phone connection. For example, coaxial cable provided by a cable television company such as Time Inc. can replace the telephone link.

AT&T SETS VIDEO STANDARDS

This *New York Times* heading for May 21, 1981 says a great deal about U.S. involvement in videotex development. On that day, AT&T specified the characteristics that it wants to see adopted for videotex systems. The announcement, made to a packed auditorium of anxious business people attending an international videotex conference, was followed by supportive statements by CBS and representatives from Canadian Telidon and French Antiope. CBS announced that it would adopt the AT&T standard for its Los Angeles videotex trials. A few months later, CBS and AT&T made public the details of a joint videotex project. The Canadian and French officials, pleased that the AT&T recommendation conformed to most to their systems, announced their intention to make the modifications necessary to achieve compatability with AT&T. The British were gloomy. Compatibility for them would mean costly changes in a commercial system already in the hands of thousands of people. One could not help but wonder where the FCC was during these events. For here was a private corporation in effect setting a national, perhaps international, policy on the future development of a mass medium. The implications of this action could well be substantial. What would consumers have to pay for videotex sets and services that conform to the Bell standard? The company's standards, which allow for the superior graphic presentation of Canadian Telidon, are certainly more advanced than those of the British. But here, as elsewhere, more advanced means more expensive. People will have to pay more for sets, equipment, and services compatible with the AT&T system. So why opt for a more expensive set of standards? One reason is that corporate executives worry that the less sophisticated British alpha-mosaic system is inadequate to reproduce advertising graphics and corporate logos. The more advanced alpha-geometric system supported by AT&T is therefore better suited to advertising. For *this* people will have to pay more? But, counter the marketplace supporters, no one will be forced to buy a videotex set and its services. AT&T may well be supporting standards that the marketplace won't bear. But what marketplace? Higher prices may mean that some will have to do without videotex; but those who can afford to will not be among them. And as videotex becomes a more and more vital tool, to retrieve basic information, to communicate with others, perhaps to work, these people will have the information advantage.[20] The FCC, at this time, does have ultimate authority to set U.S. videotex standards. Concerned people will have the opportunity to present their case to the Commission. But given the power of AT&T and the support it has received from others, notably CBS, who want to enter the videotex market, it is unlikely that the Commission can resist the Bell proposal. This is an example of the deeper meaning of deregulation. Concretely, it means that corporations whose primary interest is profit are relied on to set national policy on the production and distribution of information resources. This section explores further some of the implications of this policy by considering U.S. involvement in videotex development.

The U.S. has not been a driving force behind the development of videotex services. In part, this is the result of deregulation. Unlike France, Canada, or even Britain, where government involvement has ranged from providing some subsidies to overall management, U.S. government agencies have not been involved to any substantial extent in videotex development. Consequently, there has been very little discussion of the public service potential in videotex.[21]

This lack of government involvement has left the power to set the pace for videotex development in the U.S. to private companies. Until recently, this pace has been slow. One reason is that U.S. companies have extensive investments in videotex competitors, including videodiscs, cable television, and personal computers. Nevertheless, as Table 5 indicates, the videotex field test market in the U.S. is growing. While these trials typically involve small numbers of participants, they include a wide range of distribution channels (phone, cable, broadcast), formatting procedures (32x16, 32x20, 40x20, 80x24), display (home television, computer terminal, videotex set), and input devices (numeric keypad, alphanumeric keyboard). Consequently, the most important impact of "deregulation" is no longer on the pace of development, but is on the kinds of services that corporate development is providing. The videotex marketplace now contains the biggest names in the broadly-defined information industries.

The growth of U.S. corporate involvement in videotex is due primarily to the recognition that videotex has substantial revenue potential. Banks are coming to see that, as one *Business Week* article put it, "transaction processing in financial services appears to be the trigger application that the public would be willing to pay for."[22] Consequently, Chemical Bank is spending several million dollars to conduct a videotex banking and information service with 200 families living in a wealthy New York City suburb. Based on market survey predictions, over 20% of U.S. retail sales in 1990 will be conducted by videotex. This has prompted large retailers such as J.C. Penney and Sears to test electronic catalogs. In addition to banks and retail outlets, a long list of major publishing companies are recognizing the virtue of diversity by entering the electronic market. Much has been made of worries that newspaper companies feel about AT&T plans for an electronic yellow pages modeled somewhat after the French national project. Newspaper publishers count on advertising for almost half their revenues. Visions of AT&T siphoning off these revenues with an instant source of information on that car or vacuum cleaner you want have been the subject of numerous articles.[23] However, for all their outward expression of fear, newspaper publishers have not shied away from the videotex business. Newspaper investment in videotex surpassed the $20 million mark by the end of 1981. In fact, one of the most ambitious videotex trials in the U.S. is the product of a marriage of two would-be competitors, AT&T and Knight-Ridder, one of the nation's largest newspaper chains. Newspaper companies still fear that AT&T will one day be the dominent producer and distributor of essential information (a

TABLE 5. U.S. Companies involved in videotex market tests.

INFORMATION AND SERVICE PROVIDERS			
Finance	*Publishing*	*Retailers*	*Others*
Banc One Chemical Bank Citibank Merrill Lynch United American Bank	Dun & Bradstreet Dow Jones Harte-Hanks McGraw-Hill New York Times Readers Digest Time Times-Mirror	B. Dalton Comp-U-Card Federated Department Stores Grand Union J.C. Penney Sears, Roebuck	American Airlines AT&T Associated Press New York Stock Exchange
SYSTEM OPERATORS			
AT&T CompuServ CBS Cox Cable Communications	Dow Jones Online Computer Library Center Sammons Cable Communications	Source Telecomputing Time Times Mirror Cable	Viewdata (AT&T-Knight-Ridder) Warner Amex Cable Communications

TRANSMITTERS		
Common Carriers	*Broadcasters*	*Cable Companies*
AT&T	CBS NBC Westinghouse	Cox Cable Sammons Cable American Television & Communications Times-Mirror Cable Warner Amex Cable
HOME TERMINAL MAKERS		
Apple Computer Atari IBM	Tandy Texas Instruments	Western Electric Zenith Radio

Source: Business Week, June 29, 1981, p. 77.

prospect that prompted one newspaper lobbyist to offer this understatement: "This is one instance where deregulation won't lead to increased competition.") But newspaper publishers aren't taking any chances. Some are in fact joining with cable companies or buying their own franchises to develop alternatives to the Bell distribution network.[24]

This is little in the way of a government alternative to this corporate-controlled videotex marketplace. The only direct government presence in interactive videotex is through an agricultural information experiment—Project Green Thumb. The only other government-sponsored project is a broadcast teletext trial involving four agencies including the Corporation for Public Broadcasting. CPB has been interested for some time in developing revenue-producing services for the meagre public broadcasting system in the U.S.[25] This experiment, under the direction of the New York University Alternative Media Center, makes use of WETA, the Washington, DC public television outlet to test such potential.

In sum, U.S. companies are preparing to make videotex a major mass medium for both home and business. Since these companies are typically leaders in their other respective communications fields, countries that have gotten an earlier start on videotex can only hope that U.S. firms will adopt their advances and perhaps consider joint ventures. The U.S. is particularly important from the market end as well. With the largest mass market for electronics in the world, with the highest concentration of telephones, televisions, computers, and other telecommunications devices, the U.S. is vital to the sales plans of videotex businesses throughout the world. As the counsel to Antiope Videotex Systems, Inc. put it, "The U.S. market is crucial to whatever technology becomes the worldwide information-exchange system."[26]

Videotex in Other Countries: National Policy or Free-for-All?

A number of other countries have experimented with videotex. Most of these apply systems that we have already discussed. The most advanced technically and the closest to mass marketing are the Japanese Captain service, the West German Bildschirmtext, Viditel in the Netherlands, Telset in Finland, a Telidon system that provides government information in Caracas, Venezuela, and a Prestel variant in Switzerland.

Minor technical variations aside, the significant difference among nations now using or considering videotex is the extent of national debate and planning on videotex. Some countries, notably Sweden and Australia, have proceeded slowly with videotex, out of a concern that such a system be integrated into a national information system, that sufficient privacy protections be provided, and that they are put to the best social, as well as market, advantage. These countries are under enormous pressure from videotex promoters who hold the models of the U.S. and Britain. The latter have relied less on national debate than "market forces" to determine the place of videotex in society. Countries such as Finland and the Netherlands have succumbed to these pressures by developing systems in the hope that the presence of working systems will stimulate debate on a national information policy.[27]

Notes

[1] Richard Hooper, "The UK Scene—Teletext and Videotex," *Videotex '81*, Middlesex, UK: Online Conferences Ltd., 1981, p. 132.

[2] On the struggle over broadcast television standards see Christopher H. Sterling and John M. Kitross, *Stay Tuned: A Concise History of American Broadcasting*, Belmont, CA: Wadsworth, 1978, pp. 293-304.

[3] For an optimistic view of these issues see Efrem Sigel (Ed.), *Videotex: The Coming Revolution in Home/Office Information Retrieval*, White Plains, N.Y.: Knowledge Industry Publications, 1980.

[4] The Association also predicted that there would be 100,000 Prestel receivers in use by 1979. By October 1981, there were 13,000 such sets. See Roger Green, "Post Office Gives Viewdata a Wrong Number," *New Scientist*, October 30, 1980, pp. 300-303.

[5] Hooper, "The UK Scene. . . ," op.cit., p. 132.

[6] Brian Champness, "Social Uses of Videotex and Teletext in UK "*Videotex '81*, op.cit., p. 333. For the full study see "Teletext and Prestel: User Reactions," London: CS&P, Ltd.

[7] Hooper, "The UK Scene. . . ," op.cit., p. 132.

[8] Ibid., p. 133.

[9] Peter Head, "Prestel—From the Point of View of One Information Provider," *Videotex '81*, op.cit., pp. 138 and 140.

[10] Simon Nora and Alain Minc, *The Computerization of Society: A Report to the President of France*, Cambridge, MA: MIT Press, 1980. The word is derived from the French terms *TELEcommunications and inforMATIQUE*. It is now used in English as *telematics*.

[11] *Ibid.*, p. 2.

[12] From Telephones to Telematique," *Telecom France*, May 1981, p. 8.

[13] R. D. Bright, "The Telematique Program," *Videotex '81*, op. cit., p. 37; *Telecom France*, op. cit., p. 8.

[14] U.S., FCC, *Petition for Rulemaking In re Amendment of Part 73, Subpart E of the Rules Governing Television Broadcast Stations to Authorize Teletext*, July 29, 1980.

[15] Phil Hirsch, "France's Inteltext May Get U.S. Pilot Test in '82," *Computerworld*, October 12, 1981, p. 8.

[16] French Telematique Takes Off," *Telecom France*, May 1981, p. 20.

[17] "Can 30m French Terminals Be Wrong?," *The Economist* August 22, 1981, p. 20.

[18] Canadian Government Expands Telidon Program by $27.5 Million," *Canadian Embassy Press Release,* Washington, DC, February 6, 1981.

[19] P. G. Bowers, "The Educational Community and Videotex Applications," and J. L. Campbell & M. B. Gurstein, "Social Impact/Social Uses of Videotex in the Canadian Context," *Videotex '81*, op. cit., pp. 317-330; David Godfrey and Douglas Parkhill (Eds.), *Gutenberg Two,* (2nd revised ed.), Toronto: Porcépic Ltd., 1980; Shirley Serafini and Michael Andrieu, *The Information Revolution and its Implications for Canada,* Ottawa: Department of Communications, 1980.

[20] The proposed videotex standard is reported in American Telephone and Telegraph Company, *Videotex Standard: Presentation Level Protocol,* May 1981.

[21] For a discussion of this issue see Institution for the Future, *Teletext and Viewdata in the U.S.: A Workshop on Emerging Issues*, Menlo Park, CA: Institute for the Future, 1979.

[22] *Business Week,* June 29, 1981, p. 74.

[23] According to Katherine Graham, publisher of *The Washington Post* and President of the American Newspaper Publishers Association, "The possibility of one giant corporation becoming the information supplier. . . to four out of five American households strikes many of us as profoundly troubling." *Boston Sunday Globe*, "The New Electronic Newsboy," August 31, 1980.

[24] For a discussion of newspaper company involvement in videotex and the fear that many newspaper people have for the transformation that videotex may bring about, see John W. Alhauser (Ed.), *Electronic Home News Delivery: Journalistic and Public Policy Implications,* Bloomington, IN: Indiana University, 1981.

[25] See Sheila Mahony, et al., *Keeping PACE with the New Television: Public Television and Changing Technology*, New York: The Carnegie Corporation of New York, 1980.

[26] *Business Week,* November 17, 1980, p. 152.

[27] *Videotex '81*, op.cit., pp. 113-130 and 163-198 report on videotex developments in Japan, Latin America, and Europe, excluding France and Britain.

TRANSMISSION

PART II: PRACTICE

ERNIE KOVACS: VIDEO ARTIST

Robert Rosen

ERNIE KOVACS was commercial television's first (and some say only) video artist. From the time of his first local broadcast in Philadelphia in 1950 and for more than a decade until his untimely death in 1962, Kovacs managed—against all odds—to use television as a vehicle for exploring the outer limits of video as an aesthetic and creative medium. His style was distinctly post-modernist: a fusion of the avant gardist's concern with the formal aspects of video space and the showman's instinct for accessible mass entertainment.

Kovacs has found an honored place in the history of television and in popular memory as a wacky master comic during the "golden age" of TV comedy. His satirical take-offs on radio and film personalities are devastatingly funny and his characterizations of the Nairobi Trio, Percy Dovetonsils, and the Question Man have become cultural legacies. But beyond his already recognized talents as a comedian, Kovacs deserves contemporary acknowledgement for his accomplishments as a pioneer video experimentalist.

Kovacs was convinced that the fledgling medium could never come to realize its full potential if it were treated as little more than filmed vaudeville or illustrated radio. Rather, he sought to develop the inherent characteristics of television: the spontaneity of live broadcast, the distinctive aesthetic of the video screen, the intimacy of direct address to the home viewer and, finally, the possibilities for provoking an interactive relationship between the performer and the audience.

Kovacs preferred the rough edges of live broadcast (or an unedited kinescope) to the refinement of smoothly edited film. With nervous energy tempered by a cool exterior, Kovacs scurried from one barely rehearsed skit to another, risking at every moment losing control in front of a live audience very much like a tight-rope walker performing without a net. Edie Adams, his talented wife and creative partner, became the master of the ten-second fill—a diverting chatter that would give Ernie a breather to figure out what was happening next.

The performers were busy, but so was the audience. Using an array of strategies that, I suspect, would have pleased Bertolt Brecht, Kovacs created a dynamic interactive relationship with the home audience. Some participation was direct, as for example when viewers were invited to send gifts to Kovacs' pet flea or take "snep-shots" of the program off the tube, or to write letters that were answered over the air. More often, the interaction resulted from a style of communication that precluded audience passivity. The improvisational quality of the sketches as well as Kovacs' constant toying with the absurd and even the surreal required that viewers remain alert to participate actively in the giving and finding of meaning.

Audience self-awareness was further prompted by his preoccupation with self-reflexivity, the modernist requirement that an art medium take itself as its subject matter. Crew members were identifiable by first name; the cameras focused on one another and on the studio setting; accidental slip-ups were never hidden but rather became a part of the program. To watch Kovacs was to witness a medium in search of itself, an art form in the course of discovering its scope and limits. Kovacs proved that TV need not be just talking heads and that popular entertainment could be the cutting edge of an avant garde.

Notes on the photographs

Ernie Kovacs' special program titled *Eugene* (January, 1957) can be singled out as television's first 'no dialogue' masterwork. Analogous to Buster Keaton's classic of the silent film era, *Sherlock, Jr. (the projectionist), Eugene* is a self-reflexive work which defines the nature of the TV viewing experience while confronting the intrinsic properties of the medium. Having come from radio, as was the case with many 1950s TV performers, Kovacs was well aware of the limited use of television as merely a picture radio. Utilizing a unique array of sight gags and sets of his own design, and employing subtle sounds, music and the intimacy unique to television, Kovacs created a body of work which has had a lasting effect on TV producers; programs such as *Laugh-in* and *Saturday Night Live* are obvious examples. His influence can also be seen in the context of video performance art, most notably in the work of William Wegman.

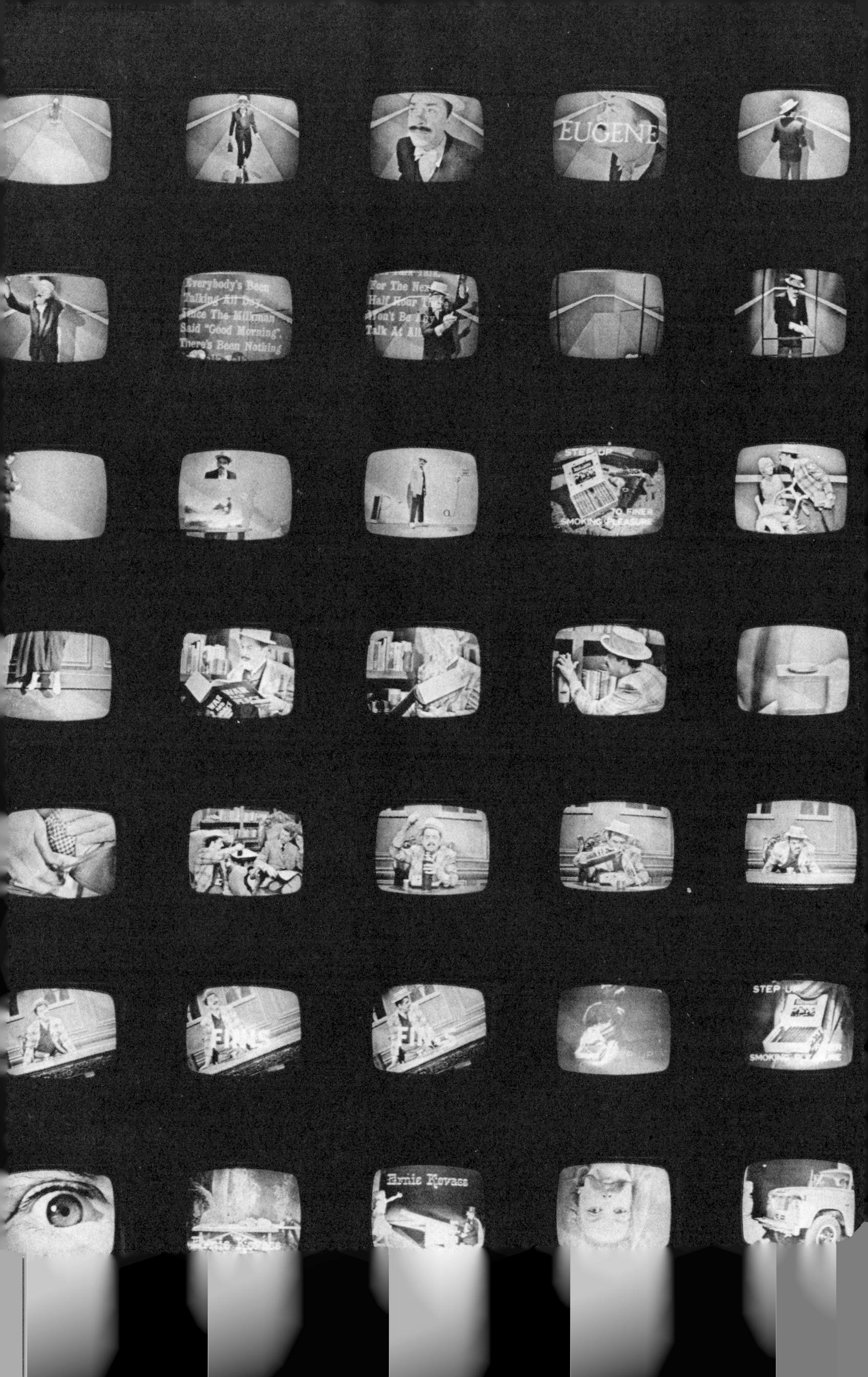
EUGENE
Everybody's Been
Talking All Day
Since The Milkman
Said "Good Morning".
There's Been Nothing
Talk At All
STEP UP
TO FINER
SMOKING PLEASURE
STEP UP
Ernie Kovacs

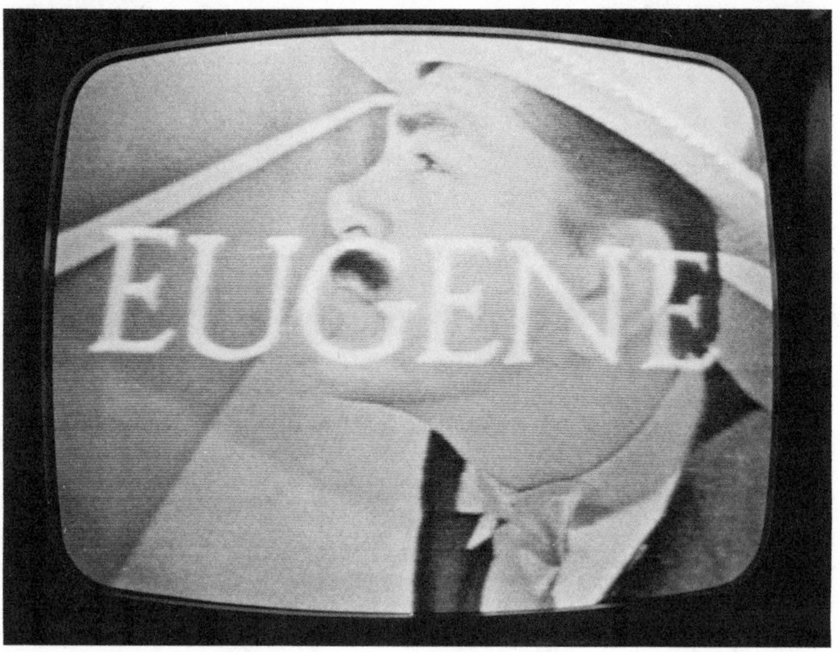
EUGENE

For The Nex
Half Hour The
on't Be
Talk At All

Everybody's Been
Talking All Day
ce The Milkman
Said "Good Morning",
There's Been Nothing

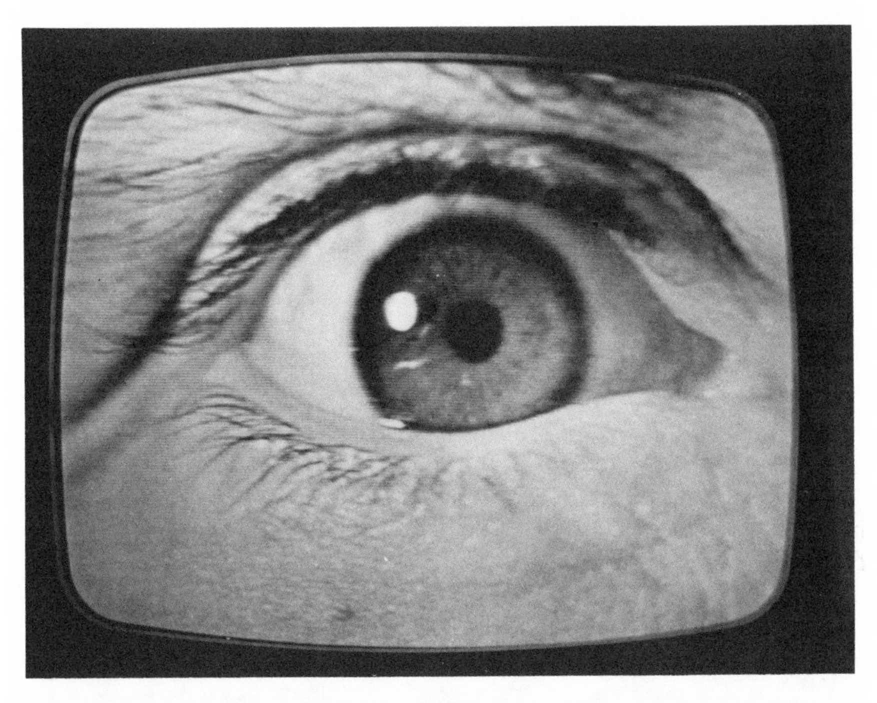

Ernie Kovacs

NAM JUNE PAIK'S VIDEOTAPES

David A. Ross

"*. . . and* TV Guide *will be as thick as the Manhattan telephone book*"
(from the introduction to *Global Groove*, 1973)

THAT LINE IS MORE THAN PROPHETIC, it is frightening. What would it mean if *TV Guide* was as thick as the Manhattan tlephone book, or to take it a step further, what if *TV Guide* was the Manhattan telephone book?

In the late sixties, the emergence of new television technologies such as cable TV and low-cost video recording seemed to promise not only a future of video abundance, but a potentially revolutionary decentralization and inversion of the virtual monopoly that corporate commercial interests held in relation to the electronic media. Artists were quick to recognize the implications of this potential shift, and while their participation in the "alternate media" movement comprised only a small (though highly visible) part of the overall activism, they did play a significant role in reforming our impression of what television was and what it could be.

Among the first of the artists who recognized the value and need for this kind of reformation was the Korean-born artist and musician Nam June Paik. Although initially interested in television from the perspective of the composer, he quickly found that the problems of music, performance, and television were closely linked by a shared set of principles. One key principle is that which links Paik to the German playwright Bertolt Brecht: the destructive nature of the one-way, passive delivery systems for electronic media. Much of Paik's TV sculpture and videotape work is directed toward the activation of the TV audience. A related principle underlying Paik's video art can be observed in his attempts to find solutions to a problem taken from the field of information science—the imbalance between information input and output with all of its physiological, political, and social ramifications.[1] As Paik saw it, before the rise of video as an art form, TV technology had been developed to make passive consumers of its audience. Paik wanted to humanize the technology by opening the role of producer to whoever had the need and/or desire to activate their relationship to television.

In 1972, Paik expressed a sentiment that seems to contradict the implied producer populism of "videology." In a letter to the editors of *Radical Software* he noted: "Currently there is a danger that video becomes like 'poetry,' . . . one guy writes, and only his immediate friends appreciate. . . . I don't know how many un-edited dull tapes I had to sit through politely. . . . We should be more conscious of the situation that we are in in the era of information overload and it means information-retrieval is more tricky than information recording."[2] This realization marks the end of Paik's first period of experimentation with television program production (1965-72), and the beginning of a period in which he produced some of the most amazing single-channel (videotapes for one monitor) television programming ever created.

"I am a poor man from a poor country, so I have to be entertaining all the time," said Paik in numerous interviews in the mid-1970s,[3] but that is not the only reason that he developed his well-known later style. A brief consideration of Paik's early videotapes may provide insight into this complex body of work, and a broader understanding of the artist's use of television as a creative medium.

Paik's involvement in television came out of his work in electronic music, a field in which the availability of a tape recorder was crucial in composing and performing. Paik "played" the television like a musical instrument. Just as in his prepared piano techniques he physically altered the piano to produce unaccustomed sounds, so in his television "sculpture" he mechanically adjusted the circuitry, tubes, condensers, and so forth, to produce unaccustomed images—to distort the "found" imagery or signals. In a way, Paik's entire video career can be seen as a highly persistent attempt to get deeper and deeper into the core of the apparatus (technically, ideologically, spiritually). At this point, his work was a clear indicator of the place artists had had in the development of television's internal grammar as well as its public image.

But before 1965—before the development of the portable video camera and recorder—there was no practical way for Paik to record his television works. His first move into the core of television came in the now mythic exhibition "Electronic TV," held at the Café à Go-Go in 1965. In October of that year, Sony had introduced the first of its portable consumer-grade video recorders to the American market. Hearing about the arrival of the first shipment in New York, Paik rushed uptown to the Liberty Music Shop and plunked down the remainder of the grant money that had been just barely supporting him in the U.S. since his arrival from Europe. His cab ride downtown was held up in a traffic jam, and upon investigation Paik learned that the cause of the jam was the motorcade bearing Pope Paul VI on his visit to New York. Paik made a brief tape of the passing parade, and showed it that evening at the café amid a flurry of proclamations including the now classic line: "As collage technic replaced oil-paint, the cathode ray tube will replace the canvas."[4] Attending those first shows were John Cage, Merce Cunningham and a few well-placed foundation people. The word was out, Paik had found a crack in a wall, and was signaling the start of television's first popular revolt. An artist had, quite simply, appropriated the tools of television production for purely aesthetic ends. Nevertheless, though Paik's video sculpture, performances, and writing of that period were extremely influential, his actual programmatic contribution to media activist causes was minimal—his videotape work was personal, experimental, and rarely exhibited.

Unlike most media activists of the late sixties and early seventies, Paik's own brand of Fluxus activity—"neo-Dada" (in Beuys' words)—was aimed at the development of a new ontology of music by exploring and expanding music's *visual* potential. Paik, in a 1972 letter to John Cage, noted: "I always think that my past 14 years is nothing but an extension of one memorable evening at Darmstadt '58."[5] What Paik told Cage was that his video work was, in effect, an extension of his notorious Fluxus concert/performance works of the late fifties and early sixties which culminated with his well-known *Étude for Pianoforte*, where he introduced the notion of physical danger into a piano work: the performance ended in an assault of audience members John Cage and David Tudor. The assault was not made out of antagonism; rather, it was a gesture designed to involve the otherwise passive audience members—of which Cage and Tudor were members. But the reference to Darmstadt in Paik's 1972 letter also recalls an earlier work, *Poly-hetero-phony*, a musical collage in which he introduced into his own work the notions of randomness and indeterminacy and the intermingling of Eastern and Western thought. This form of intellectual and spiritual entanglement, culled from Cage and the lessons of the Dadaists, has provided the matrix for Paik's videotape work. Within this matrix he has been able to address a range of concerns that surface and resurface, including the paradoxes of time perceived and time understood from both the linear Western perspective and the non-linear Eastern one, the hegemony of European academicism in art and music, and of course the role of the revolution in communications and control (cybernetics) in the transformation of global consciousness.

Present in Paik's early videotape works are many of the elements that, after 1972, would form his distinctive style. But what remains clear is that the early works reflect the same anger and frustrations at audience passivity that surfaced in the *Étude for Pianoforte*, but now focused upon the technology and ideology of television.

In *Variations of Johnny Carson vs. Charlotte Moorman* (1966), Paik produced his first videotape equivalent to his prepared television sets. The content of the tape, though wonderful and hilarious in its own right as a time-capsule piece in which Johnny does a series of double-takes at the "whacky" avant-garde, is standard Carson fare. Paik taped the show off the monitor screen and then placed a live wire across the reel of recorded material. The live wire acted as an electromagnet, erasing the material directly beneath the wire, so that what appears when the tape is replayed is the Carson-Moorman interview with a brief erasure at every four seconds, but with increasing frequency as the tape nears the core of the reel. We are confronted with a work in which the artist literally reached into the program (albeit after the fact and in the privacy of his studio) and marked his presence, forcing a recognition that something had changed, that the order had been tampered with.

A different work from the same period, *Variations on George Ball on Meet the Press*(1967) can be seen as another exercise of literally inserting the artist's hand into the television process. An off-screen tape of former Under-Secretary of State George Ball (who had recently resigned his Johnson Administration post, ostensibly in moral opposition to the war in Vietnam) is re-recorded in a mechanical-transfer process which allowed Paik to manipulate the take-up reel by hand, at irregular intervals. By controlling the tape speed in this way, Paik produced not only pictorial distortions, but also broke the real-time fidelity of video recording, demonstrating how very illusory the concept of "real time" is on videotape. The work can also be read on a political level: Paik's choice of Ball as the subject of the work underscores the way in which normal media coverage distorts and manipulates its audience and its subject. Paik had learned that the media's emphasis on the moral grounds for Ball's resignation was misdirected—Ball had in fact resigned because he felt that the war effort was not cost effective. This was the same year, after all, in which Marshall McLuhan's theories were first reaching the mass audience, creating a climate for the questioning of basic assumptions about mass media. It was also the same year in which Paik created *McLuhan Caged*, an electromechanical distortion of McLuhan's videotaped face in the landmark Museum of Modern Art exhibition "The Machine."

The next period in this early stage of Paik's video career saw both the development of his instrument, the video synthesizer (a device which mixed distorted live-camera signals with pure electronic signals[6]), and the building with his productive relationship with public television. In 1968 Fred Barzyk, the pioneering producer at WGBH-TV, Boston, invited Paik, along with a group of visual artists—among them Alan Kaprow, Aldo Tambellini, and James Seawright—and others, to participate in a project underwritten by the short-lived Public Broadcasting Laboratory. Paik's segment, *Electronic Opera No. 1*, in *The Medium is the Medium* was his first broadcast piece. A video collage, it consisted of short bits including (appropriately for the time) Nixon's distorted face juxtaposed with a psychedelic nude go-go dancer, electronically generated Lissajous figures and, finally, Paik's call for interactive two-way television. "This is participation TV," says the voice-over as a particularly beautiful segment proceeds visually; "please follow instructions." Paik's voice then directs the viewer to "close your eyes," and then, after five seconds, to "three-quarter open your eyes," and so on. He ends the piece by instructing the audience to turn off their sets. The piece demands to be viewed by offering a fast-paced mix of nudity, low comedy, and electronic imagery, and then, literally "demands" to be viewed partially and finally not viewed at all. The work functions as more than just satire, since it leaves the viewer with the question of who actually is in control: what is the basis of the decision to become a passive receiver of TV—or of art, for that matter?

After the success of *The Medium is the Medium*, Paik convinced WGBH to build the first model of his video synthesizer, the concept of which was based upon his experience with electronic music as well as his growing understanding that the kind of time an artist needs to produce work is not available within the extraordinarily costly allotments of studio and engineering time an artist can expect from even the most generous producer.

Paik described the frustration in an article for the second issue of *Radical Software*: "(I use technology in order to hate it more properly). . . . In the heated atmosphere of the control room, I yearn for the solitude of a Franz Schubert, humming a new song in the unheated attics of Vienna. . .
Ironically a huge machine (WGBH, Boston) helped me to create my anti-machine machine."[7]

Around the same time, Paik agreed to help develop the Video Department for the newly opened California Institute of the Arts (building with his collaborator Shuya Abe a synthesizer for them as well), agreed to build a Paik-Abe synthesizer for WNET-TV in New York, and built one for the Experimental Television Center in Binghamton. With the field well sown with instruments, Paik, and a growing band of students and followers, began to learn to play with the new tool.

Paik, unlike his followers (and fellow inventors), immediately made his instrument available to the public, in exhibitions to the Galeria Bonino and elsewhere. More important, though a strong advocate of electronic imagery, Paik never lost sight of the fact that his collage format—the juxtaposition or layering of images, viewpoints, and spaces—was essential to avoid the production of what he termed "vuzak." However, his own production attitudes notwithstanding, Paik described the synthesizer in hyperbolic yet poetic style: it would

enable us to shape the TV screen canvas
as precisely as Leonardo
as freely as Picasso
as colorfully as Renoir
as profoundly as Mondrian
as violently as Pollock and
as lyrically as Jasper Johns

Paik even went so far as to say that the synthesizer could create a TV-tranquilizer, which he described as "an avant-garde artwork in its own right."

In fact, the Paik-Abe synthesizer is by contemporary standards a relatively crude analog device, lacking the computers and digital controls most broadcast industry special-effects generators use today to produce a far more extraordinary range of optical effects. What Paik's analog synthesizer represented, however, was the fact that artists could take that next step into the core of television by reinventing the tools of production to fit their own needs. This, and the fact that it made possible the production of non-representational television imagery, must rank Paik and Abe's device as one of the most significant works of sculpture of the last twenty years. (As in Paik's earlier television sculpture, the synthesizer is seen in the same space as the images it produces, so that the instrument becomes part of the visual experience of the piece.)

Paik's next WGBH project was a four-hour "live" broadcast, *Video Commune* (1970), in which he, along with the WGBH staff and people from off the street, manipulated the synthesizer to a rock-and-roll soundtrack creating, to quote Paik, not "cybernated art . . . but art for cybernated life."[9]

Later that year, Paik was asked to participate in a TV program featuring video artists working in collaboration with the Boston Symphony Orchestra. In his prerecorded segment for this special broadcast, Paik illustrated the third movement of Beethoven's *Piano Concerto* by subjecting the score to a full-scale broadside attack, with tongue in cheek, of course. Mixed into a swirling montage of video-synthesizer imagery, electronic distortions and colorized film, we see a bust of Beethoven, as familiar as the Beethoven passage itself, transformed by colorizing, twisted by the synthesizer, and finally slapped back and forth by a hand that enters the frame ostensibly to slap some sense into Beethoven (or perhaps the audience). What is the sound of one hand slapping Beethoven, one wonders, as a (toy) piano burns to the final strains of the piece.

In the sense that it again represents Paik's Fluxus rejection of the hegemony of European art, and his penchant for low comedy with serious overtones, the BSO piece set the tone for most of the videotapes that he would produce in the seventies. With the exception of the *Tribute to John Cage*, the BSO piece was also Paik's last at WGBH, as he received an invitation to participate in WNET's newly formed TV Lab in New York.

Paik's first tapes produced at WNET were experiments in which, among other things, he tried to develop a video equivalent to his profound sculptural innovation, the *TV Bra for Living Sculpture*(1969), worn by none other than his indefatigable performer and collaborator, Charlotte Moorman. In these experiments, Paik tried to replace the actual televisions in the *TV Bra* with superimposed, chroma-keyed images. The sculptural notion of the *TV Bra*, and its video counterpart, both serves as a metaphor about the place of TV in our culture and provides a radically new context in which to consider TV.

The performances (with the working title *TV Experiments*) were never aired, but they did find their way into many future Paik productions. As a matter of fact, this element of Paik's style must be given proper notice. Paik's collage style, as it evolved, is closer to a video compost, with no old work ever discarded, and nearly every bit or routine resurfacing to give the viewer the comfortable feeling that this seemingly foreign material is somehow familiar. There is also something here that relates again to Paik's preoccupation with time, for as the artist sees it, video not only confers immortality upon its subjects, but is itself the essence of immortality. "Once on videotape," Paik has been known to say, "you are not allowed to die."

The Selling of New York was a series of short, emblematic segments that Paik produced for insertion into the late-night schedule of WNET-TV. This highly entertaining sequence of sketches comprised a veiled critique of the disparity between the marketing of New York as a media center and idea factory and the reality of the lives lived by New Yorkers themselves. The media-hypester, played by Russell Connor, drones on in monologue, citing New York's relative number of television homes as compared to Missoula, Montana, and other thrilling facts and figures. In contrast, "real" people are seen trying to ignore the ubiquitous Connor monologue, which seems to be playing on TV sets everywhere. In one of the highlights, Connor (representing the unrelenting TV presence) carries on about the statistical reduction in crime while a thief unplugs and steals the set on which the show is playing. In this group of Ernie Kovacs-inspired blackout routines Paik inserted Japanese television commercials for Pepsi and a Tokyo fashion concern, both of which appropriate "Americanisms" into the fabric of their selling style. The point about New York's selling is well made. Paik once remarked that the earliest childhood dream image he could recall was the face of Shirley Temple, which had been widely marketed as the image of America when Paik was growing up in Korea. Current media critics allied to the UNESCO New World Information Order, for example, note with anger that the flow of information was decidedly one-way—from the industrial world to the media-subjugated developing world. *The Selling of New York* makes this point in the understated way in which all of Paik's political philosophy is delivered.[10]

But for all the pointed humor of *The Selling of New York*, the fractured nature of the work's presentation did not really provide Paik with the forum or form he needed. His first real opportunity came in his production *Global Groove* (1973). Not only did *Global Groove* allow him to create a vehicle for the short bits he had produced, but it also allowed him to expand the public audience for video art while acknowledging the contributions of his friends and colleagues. This work set the benchmark for a generation of aspiring video artists in its state-of-the-art mix of entertainment values with a rigorous adherence to Paik's own aesthetic.

Global Groove is many things. It is, for example, the model program guide for the time when *TV Guide* will not only be "as thick as the Manhattan telephone book," but—with everyone becoming a TV producer—will *be* the Manhattan telephone book. Developed from the 1965 essay in which he predicted a twenty-four-hour schedule for his "Utopian Laser TV Station," *Global Groove* features Charlotte Moorman playing the electronic video cello, Allen Ginsberg chanting in glorious synthesized color, Mitch Ryder and the Detroit Wheels' "Devil with a Blue Dress On," music by Karlheinz Stockhausen, John Cage providing anecdotal relief, films by Robert Breer and Jud Yalkut, and an assortment of recycled bits from earlier experiments and programs. It is a prime example of Paik's use of three distinct segment-types to achieve a balance, a balance Paik describes as one between three corresponding experience levels that constitute stable human life. Normal wakefulness is represented by real time—representational imagery—while the sleep and the dream state are represented by less rational constructions often involving compressed or extended temporal rhythms. These two states, which account for the majority of the work/life-cycle, are supplemented by the third state, the heightened levels of consciousness associated with all forms of ecstasy, induced physically, chemically, spiritually or however. These segments are generally the hyperkinetic, synthesized and colorized bits that seem to emerge from the other two types of segments in seemingly random patterns. This fast-paced, densely edited work was Paik's first use of state-of-the-art editing techniques, and represents a real milestone in the tempering of technology by precise aesthetic purpose. "I make technology look ridiculous," Paik has said, but in this work the artist revels in the spirit of the new technologies in a way which belies his oft-stated belief.

In a 1975 interview with *New Yorker* critic Calvin Tomkins, Paik stated that he "has no scruples," and as such recognizes the need to be entertaining all the time. But Paik's next comment reveals his true ambivalence about television entertainment: he noted that he would rather make boring programs because television is already so filled with interesting fare; that he would rather corrupt himself than repeat his earlier "sublime" works.[11] It becomes evident that Paik relishes the contradictions inherent in the very idea of the interface of an uncompromised aesthetic and politic within a context of total compromise that *is* broadcast TV. But it also appears that there exists some level of discomfort in Paik's reconciliation of his videotape work with his fully uncompromised performance, installation, and written works.

This conflict and its complementary resolve were never more evident than in his tour-de-force appearance on the now defunct but then highly rated Tom Snyder show *Tomorrow* in 1975. Paik's segment on the show (a special on video art in typical Snyder style)[12] was a live remote from Paik's Mercer Street loft. In his performance, he demonstrated that the philosophical context of commercial television was no longer a barrier to his control of or his approach to the medium. Though the context of Paik's performance/demonstration was not all that different from a standard TV visit to the home of a celebrity (Jackie Kennedy at the White House comes to mind), Paik managed to turn the event into an outrageous Fluxus performance. Rather that let Snyder obtain or retain control, Paik immediately began asking the questions of Snyder, turning him into an unwitting subject. He asked his "host" whether TV was an entertainment or a communication medium, and after praising Snyder for the sincerity of his response ("information"), he began reading from a book (Don Luce and John Sommer, *Vietnam: The Unheard Voices*, 1969) in which the preface states that the American failures in Vietnam were essentially the result of failures of communication and understanding. Paik then introduced the *TV Buddha*, and quickly moved to the *TV Chair*, in which the television screen is given the analogous place of a toilet bowl receptacle. On the set was Snyder's face (prerecorded by Paik off the air). Paik sat down on top of the chair while the "live" Snyder squirmed uncomfortably with slow-dawning recognition of the symbolism of Paik's interactive performance. A flustered Snyder tried to end the segment, but Paik kept control of the conversation by insisting upon introducing Shigeto Kubota and her work, using the network stage for an appropriate plug for her career, not unlike Charo plugging her next Vegas appearance.

Although the *Tomorrow* show sequence had no distribution as a tape, it has all the features of a Paik work of the period, from Ginsberg chanting to Charlotte Moorman in TV cello space; it is in some ways the most revealing example of the interaction of Paik's style with the world of "real TV," a style that turns the context of television into a self-defeating proposition.

In 1973, Paik had finished his hour-long *A Tribute to John Cage*, an engrossing study of the strongest single influence on Paik and a true homage to Cage's unique genius. Included in the tape is a wonderful monologue about Cage by David Tudor (explaining how Cage taught him to use his stutter as a sound no better or worse than any other), a series of anecdotes by Cage (some of which had been included in *Global Groove*, of course) taken primarily from Cage's compilation *Silence*, and a performance of *4'33"* staged for the camera by Cage in Harvard Square.

But the homage to Cage had left a very important influence in Paik's ongoing career unacknowledged. And so he and Shigeto Kubota set out to produce a collaborative work with Merce Cunningham (and his video collaborator Charles Atlas) which would address both Paik's debt to the great choreographer and their collective debt to Marcel Duchamp. The 1978 work, *Merce by Merce by Paik*, is probably one of Paik's most underrated, but should be seen as one of his most direct and profound.

The tape consists of two distinct parts, the first being a suite of short dance/video pieces produced by Cunningham and Atlas. What quickly emerges is the notion that time is the subject of this work, time as experienced by the dancer in action, and the relative nature of time as the malleable component of video art. In the ending of the Cunningham-Atlas segment, we hear an off-screen woman's voice ask Cunningham, "Can art kill time or occupy it?" "No," answers Cunningham's voice as his image dances in the sky on the screen, "it is the other way around."

Paik and Kubota's half of the tape begins at this point. Their work, subtitled *Merce and Marcel*, maps the territory opened initially for artists by Duchamp's profound philosophical and spiritual insights into the nature of art; insights opened for dance by Cunningham's deep understanding of that nature.

Following the essentially unanswered question that ends the first part of the tape, Paik questions the relationship of dance to time, and asks, "Can artists reverse time?" Perhaps more significantly, can the artist, in the documentary process, defeat the implications of time's relentless progress? As Paik wrote to Ira Schneider and Beryl Korot, "Paul Valery or so said that there are only two poles in poetry . . . (abstract and semantic). . .

Do you think, that we . . . found one more pole . . . and with all new manipulatve possibility in the time-parameter of video tape . . . did we find a new pole in TIME besides the one way flow of time?"[13]

The work proceeds with a rare television interview with Duchamp by Russell Connor, edited in steps so that a Duchamp statement is repeated in rapid succession with a slight progression in each repeat. Time, in this instance, is not completely reversible, but moves forward in a staggered, talking two-step. Later in the tape, Connor is seen interviewing Cunningham in 1976, and that interview in intercut with and superimposed over the twenty-year-old Duchamp-Connor conversation. Time reversible and made manifest; video editing as the dance of time? These concepts are alluded to both in the content of the discussions used and in the manner in which the segments are combined, producing a synergistic effect, aimed at approaching the unapproachable: a definition of the role of time in art. The piece, which uses Kubota's well-known videotape homage to the grave of Duchamp, outside of Paris, continues with a sequence in which Duchamp states that repetition is a preparation for accepting the idea of death and that there is no solution because there is no problem. The tape ends with Duchamp recanting, however, giving the work a final ironic twist, stating that he preferred Man Ray's approach to the issue: "There is no problem, there is only solution."

When viewed as pure "solution" rather than as an artist's approach to problem solving, Paik's videotapes emerge in a different light. Prior to the completion of the Cunningham tape, for example, Paik finished his first version of *Guadalcanal Requiem* (1977), a work which must rank as Paik's most dense and most complexly constructed tape.[14] The tape functioned as a further exploration of time (and memory) as well as an evocative and potent pacifist artwork.

Set on the site of one of the bloodiest battles of World War II, this video performance collage once again features the bits from standard Paik-Moorman repertoire: Charlotte Moorman performing a variety of "new cello" works (including the Beuys felt-cello piece, Paik's G.I.-cello-crawl piece, and bomb-cello piece), all looking completely fresh in the context of an eerily preserved battle-scarred landscape. But the performances are set into interviews with a visiting U.S. Marine veteran of Guadalcanal, a Solomon Islander who remembers having rescued John F. Kennedy (he shows the camera his P.T. 109 tie clip), and various Islanders who are involved in the move toward independence (from the U.S.) for the Solomons. And, as the horrors of the battle are remembered, and intensified through the same kind of time-stepped editing and reiteration encountered in *Merce and Marcel*, the surreality of what Paik describes as the first war of the energy crisis takes hold of the viewer, who can't help but question the way in which war policy is rationalized historically. Though not a revisionist historian, Paik insists on his role as a tragic observer of global conflict seen (again) as a primary result of the lack of communication between divergent cultures and ideologies.

If the war and the continuing American presence in the Solomons run through the piece as undercurrents, time's passage and attendant illusions must be seen as the work's major theme. The use of a highly refined computer-editing system allowed Paik to combine video synthesis of old war film footage with the performance and interview material in such an incredibly dense manner that one is subliminally bombarded with information in rhythmically shifting time perspectives, leaving one with a feeling of video exhaustion: information overload as the battle fatigue of the future.

Perhaps this contemporary cultural malaise lies at the core of all of Paik's videotape work: the effects of information overload on an already alienated population. In a recent conversation, Frank Gillette (the video artist and photographer who in 1969 wryly termed Paik the George Washington of video) described Paik's video style in a manner which seems to address the essence of his art: "Paik's style consists in a certain force of assertions suspended between the sublime and the comic. Its standards of precision, its Buddhistic negation of will, its absence of stipulations; all combine with persuasive terms to effect an aesthetic discharge at the nausea of absurdity."[15]

Does Paik's videotape work (combined with his other activities) constitute this "antidote," or is his art merely a finger pointing at the moon? Well, as Paik has said, "the moon *is* after all the first TV."[16]

Notes

[1] Nam June Paik, "Communication-Art," in *Douglas Davis: events drawings objects videotapes*, ed. David Ross (Syracuse, N.Y.: Everson Museum of Art, 1972), pp. 11-12.

[2] Nam June Paik, "Binghamton Letter" (January 8, 1972), in *Nam June Paik: Videa 'n' Videology 1959-1973*, exhibition catalogue (Syracuse, N.Y.: Everson Museum of Art, 1974), p. 69. (Since this book lacks numerical pagination, for the reader's convenience page numbers have been assigned, page 1 being the Foreward.)

[3] Nam June Paik, interview with Russell Connor and Calvin Tomkins, in "Nam June Paik: Edited for Television," WNET-TV, 1975.

[4] Nam June Paik, "Electronic Video Recorder," in *Videa 'n' Videology*, p. 11.

[5] Nam June Paik, "Letter to John Cage," in *Videa 'n' Videology*, p. 64; in this letter, Paik pleads with the composer to take television seriously, saying that "TV is also a form of giving away . . . even more so than music."

[6] The effect of the video synthesizer relied heavily on the fact that a picture signal has one unique characteristic. Unlike audio feedback (the ear-piercing squeal heard when a live microphone and loudspeaker form a closed circuit and begin to oscillate at a high frequency), video feedback occurs when a live camera is focused on its monitor, producing a mandala-like swirl of exploding light patterns that seem to rotate around a central axis. The Paik-Abe synthesizer used this technique in combination with aspects of the video picture

controllable by simple voltage regulation—color intensity and hue, for example—to produce highly original (though ultimately predictable) imagery.

[7] Nam June Paik, "Video Synthesizer Plus," *Radical Software*, 1, no.2 (Fall 1970), p. 25.

[8] Nam June Paik, "Versatile Color TV Synthesizer," in *Videa 'n' Videology*, p. 55.

[9] Nam June Paik, "Utopian Laser TV Station," in *Videa 'n' Videology*, p. 17.

[10] *The Selling of New York* was re-edited into a longer program entitled *Suite 212*, where the "selling" sequences appeared with additional material by Douglas Davis, Ed Emshwiller, Jud Yalkut, and others, much of which Paik synthesized for inclusion in the piece.

[11] "Nam June Paik: Edited for Television," WNET-TV, 1975.

[12] Also appearing on the show were Douglas Davis and this writer; other tapes featured were by William Wegman and Peter Campus.

[13] Nam June Paik, "Letter to Ira Schneider and Beryl Korot," in *Videa 'n' Videology*, p. 63.

[14] The original version, completed in 1976, was fifty minutes long; a second version, released in 1979, was trimmed to twenty-nine minutes.

[15] Frank Gillette, conversation with the author, November 1981.

[16] Press release, Gallery René Block, New York, 1978.

SAMUEL BECKETT'S *GHOST TRIO*

Peter Gidal

GHOST TRIO was written expressly for television by Samuel Beckett in 1976 and recorded by the BBC in October of that year. It was published in the United States (Grove Press) and in England (Faber and Faber). The British publication, which first appeared in the Journal of Beckett Studies for Winter 1976, includes revisions made during the broadcast recording; hence, the British texts are, finally, based on both the writing and the telerecording. As *Ghost Trio* was written as a teleplay, it has not been otherwise performed, and will not be. The Male Figure is played by Ronald Pickup, the Female Voice is played by Billie Whitelaw. No name is given for the boy who plays the boy, in any printed text. The production was directed by Donald McWhinnie, under the constant supervision of Beckett, who also directed a German television production of the play. A trilingual publication of this, and various other recent pieces, was brought out by Suhrkamp, in Frankfurt, as *Stücke und Bruchstücke*; that edition also includes Beckett's work notes and instructions for various performances, outlines, drawings or precise movement structures, etc.—all in all, a more useful publication.

As to *Ghost Trio*, a conventional description to start: The play itself is conceived within three frames, each containing another frame and another world like a Chinese box. The outer frame is that of the television screen itself, the rectangular shape of which gives the play its most striking visual aspect other than the human figure. The voice of the woman announcer operates on this plane, explaining what is seen in the second frame, that of the visual narrative. The third plane is interior, that of the inside of the mind of the protagonist and of the music that comes from a small cassette recorder, Beethoven's "Ghost Trio," Op. 70, no. 1. The announcer's voice tells the viewer that hers is a faint voice that will not be lowered or raised and that the predominant color of their picture should be shades of gray. She describes three of the rectangular objects that can be seen, a door, a window and a pallet or simple bed on the floor, all of which are the same rectangular shape. . . . There are also three other rectangular objects not described, the cassette recorder, the stool on which the man sits, and a mirror. . . . There are also three other rectangular shapes shown and mentioned: a section of floor, a section of wall, and another view of the pallet, this time resting against the wall. . . . The narrator also describes the sole inhabitant of the room, a man. The man is immersed in listening to three musicians playing a trio, the sound emanating from inside a cassette recorder, and he is observed from outside the television set by the television audience and by the narrator, . . . he is also listening for another sound, for "her" to come. What he sees, after twice opening the door and window, and looking in the mirror, is a surprise to him. . . . It is the boy who comes, dripping from the rain, who indicates, smiling and negative, and retires down the passageway. . . we are told by the narrator that the light never changes, never dims, that the room contains no shadow.[1]

In quoting the description, I have left out as much symbolic interpretation as possible, although exegetes of Beckett's work always thrive on seeing his work as some kind of pretext, autobiographical or otherwise, as if the work were there to illustrate some notion of Beckett about his personal relation to rooms, knocking, boys, mirrors, doors, tape recorder, waiting, the number 3, and so on. Also, an intense humanism is always inferred (or interfered), which may indulge whatever writer is writing, but which has nothing to do with the work at hand. As if saying good things about the human condition were in any way akin to political action, for example, or as if a written or performed work were not a *work*, with such and such relations of production, of meaning (and meaninglessness), as opposed to some sort of preexistent truth that is somehow given back, reproduced (adequately) to be consumed into "life." Thus, there is a constant search for (a) symbol, ironically "forgetting" Beckett's often quoted statements to the effect that if he had wanted to say such and such, he would have.

I am not, here, attempting to use the author (authority) for verification. It just happens that his theoretical position for his texts is more advanced than that of most critic-academics. Hence the many attempts in *his* speech, on the subject of his work, *not* to see it as imbricated with symbolism, anthropomorphism and, ultimately, a message of humanism. A common problem: "Kafka's radical critique of humanism was one of the things that made him unintelligible to a Marxist author like Lukacs. When it comes to revealing the ties between Humanism and upper-class ideology, the Marxist demystifiers are as timid as Freud."[2] Of course, the Marxists referred to are precisely those who *cannot* cope with humanism and its demystification, whereas writings by, for instance, Louis Althusser advance a materialist critique of humanism.[3]

On with *Ghost Trio*. In most works by Beckett, the subject is charged with hyperconsciousness, the subject being the (heard) narrator, or the (seen or heard) protagonist, or the viewer in her/his identification with one or another of the former. In *Ghost Trio*—which had originally been entitled *TRYST*—to repeat, an act is narrated by a woman's voice. Through being narrated while the enactment is produced, the act, whatever it is, intially undermines any view of itself that would allow for a naturalism or a suspension of disbelief. The fact, in other words, of a narrator overtly narrating a series of actions renders problematic, at this first stage, certain identifications into the imaginary space and time of those actions. When the scene begins with "Mine is a faint voice. Kindly tune accordingly," we are given to believe in one fiction only: not the fiction of the enacted, but the fiction of the enacting, the voice of the woman speaking constituting that enacting. The *words*, thus, are the reality of the scene—the words, her words, the signifiers. Hence a convention within signification—within a play, a film, a television play, and so on—is countered from the start. The images—that series of actions seen conventionally as the necessary producers of meaning—are here given as *products*, as already determinate effects. The imaginary management of theatrical space, which represses the causality of the enactments, is withheld—not effaced or deconstructed, but withheld. The narrator's voice is presented: "It will not be raised or lowered. . ."

The script of the television play breaks down into three parts: (1) "Preaction"; (2) "Action"; (3) "Reaction." The sole characters are the Female Voice, Male Figure and the Small Boy. The music heard is from the largo of Beethoven's fifth piano trio ("The Ghost").

There is a pause after "Good evening. Mine is a faint voice. Kindly tune accordingly," after which, "Good evening. Mine is a faint voice. Kindly tune accordingly." Pause. "Look." Long pause. "The familiar chamber." Pause. "At the far end, a window." The description continues.

At some point: "Now look closer"; and the image cuts to a rigidly formalistic, frontal high-angle view of a rectangle that supposedly represents "the floor" or the "the wall." In these closer views ("look closer") we are given a less naturalistic view, one *not* coinciding with what we might have approximated, if based on inferences from the longshot of the space, which is the first shot we see—a room with a bed and a door and a window and a stool and a man sitting. Here, thus, is a second disjunction of the real or the imagined real.

As the camera moves in on the seated figure, music slowly fades in, getting louder as the camera gets closer; as the camera recedes, the music fades out; then, in silence, the camera keeps receding, until it gets back where it started, in frontal medium-longshot. The question arises as to the "nature" of the protagonist, the antagonist. This *agonist*, a present absence: is the camera a metaphor for a present absence? The very existence of a mobile unit to change the reproduction of a viewed scene ineluctably recalls another figure, a viewer, and is thus an anthropomorphism. And our own positioning through that becomes problematically inscribed through the making present—the foregrounding—of precisely such camera movements into a scene "out there." In other words, since the camera movement efface itself as the method of producing a certain kind of narrative seamlessness or narrativization for the viewer, the camera does give itself a durational hold upon the scene. This camera movement blocks that flow, *making difficult* a "natural viewing" that would see nothing but the effect—the final play, an effect without cause, magically produced. The camera movement does not hold itself, unquestioned as to meaning, or mode of production, or ideological desire toward certain ends. . . . It does not efface itself toward repressing the knowledge of how a material is worked through to produce such-and-such effects. So we are left in and unavoidably problematic area, whether or not the camera becomes a protagonist or agonist or a mechanism of differentiating from the scene as given, and no more than that—merely another function in the mechanism that produces this videoplay . This problematic area must be borne in mind as an entanglement that cannot simply be left to its power as possible and obvious metaphor. That would be the too unproblematic consumption, one which the text *Ghost Trio* in the particularity of its production as a television play denies, although metaphorical/symbolic readings/interpretations always come to hand quickest precisely because, being merely substitutive, they slip into their pregiven meanings most easily and exclude labor processes from their construction. Such meanings are, then, never given as *effects*; they are given precisely as *pregiven*. So everything stays as it is, and we have a metaphysic of absent presence, in the case of the camera described above, which would cover all blockings . . . and hold them together in the imaginary unity of the natural—in whatever style the "natural" for each specific production might be defined.

"He will now think he hears her," says V (Voice). *We* are thereby positioned in an insecure perception, fraught with the will for knowledge: who is "her"? Thus we perceive (aurally) that we do *not* know who this she is. The disembodied voice is the voice of *a* "her," a woman, but is it the her referred to? Then, also is it the voice of a her at all—of a woman, a body, a figure, present or not—or is it simply a *voice*, a different kind of character maintained from the usual one necessitating a figure? The voice that is articulated, outside of this television play as well as, possibly, "inside" it, is after all, not necessarily a part or a fragment. It may, in fact, not be disembodied; nor does it have to be embodiment of a figure, contracted into a metaphoric voice. It may be voice as articulating a specific set of words, with certain meanings—in no way a voice *of*, or *from*, or *for*. Voice may thus be "prehistorical" and in the position of forming its history at the moment of articulation, a flow of the very material of the voice, a history rememorating itself, but only from the now, the point, *now*, of inception. The material history of the voice is defined only from the moment it exists in its specific determination as voice, and not carrying in any way the baggage of poetics, resonances of theater art, or life or whatever. V: "No one." "Again." Voice directs the protagonist's actions or listening for her whom he thinks he hears, whom we hear.

When the actor "raises head sharply, turns still crouched to door, fleeting face" (Beckett's instructions in the script), he is operating within the gesture (*die Geste*) in the strictly Brechtian sense. This by now overly familiar notion has hardly been sufficiently realized theatrically to warrant its being designated a cliché. The rigid freeze of an action into a pose, which, as unposed, relapses into movement (towards), then reholds itself into the gesture ("fleeting face, tense pose . . . 'again' . . . raises head sharply, turns still crouched to door, fleeting face . . ."): such gesture relapses into movement and then reholds itself as gesture. That distancing is not evidence of a mechanistic understanding of Brechtian theory as meaning merely a distance that is only to be overcome afterward by reentry into the imaginary transparent reality of whatever fiction or pseudo-documentary is at hand. That distancing is of a truly Brechtian type often unacknowledged by Brecht in his lesser writings and in most Brecht productions. It has been affirmed, however, in the present Beckett piece as well as in the production of *Waiting for Godot* that Samuel Beckett directed at the Schiller Theater in Berlin in early 1975, with Estragon played by Horst Bollman, Vladimir by Stefan Wigger, Lucky by Klaus Herm, and Pozzo by Carl Raddatz, and which played in London in 1976, and in the same year in his production of *Footfalls* with Billie Whitelaw. Work on the relation between such Brechtian technique and the work of the Bavarian comic Karl Valentin has yet to be undertaken. Brecht started in Valentin's troupe, and worked closely with him on a co-directed film *Mysterian eines Friesiersalons* (1923). He also credited Valentin with the "invention" of the distancing technique, in his *Arbeitsjournale* and in often quoted notes on the theater. Some of the dialogues in *Godot* sound straight out of Valentin (which is no insult to either). Beckett had seen Valentin perform in 1938: "Yes, I saw K.V. in a shabby Café-theatre outside Munich. Evil days for him. I was very moved."[4]

The window of the set is not a glass pane; it is a painted-gray "window." But the bench can be sat on, the door can be opened, the pallet (a matress on the floor) can be lain upon, the cassette recorder can be listened to, the mirror can be looked into.

As the camera moves in the sitting figure, a cassette in his hand is unrecognizable at first. The camera stops, pauses, starts moving in further, with the sound faint, then louder, with the moving closer of the camera. Then the camera recedes with it. "He will now again think he hears her." The camera stays out at the original, unmoving position, and yet the sound comes on as the man (F: Figure) listens. He lifts the cassette closer to his face: he and we hear the sound—the music—at normal television level. The script states: "faint. It grows louder." In any case, the camera this time has *not* moved in, but the music nevertheless comes on, and F and we hear it. V: "Stop." Music stops. V: "Repeat." Sound comes on; image cuts to near shot; and *moving camera equals louder sound.* F raises his head sharply, as if hearing something: at that instant the sound stops. Head back into opening pose, sound returning with it. Thus the Figure controls the sound now—an effect without cause, but *given to be seen by us as without cause.* Thus we are not given an effect, but some image magically produced and, therefore, an idealism, without cause. We are given the causelessness itself. Just prior to this sequence *Voice* controlled the sound with its (her) "Stop."

In viewing this durational section, and in allowing for the contradictions of the controlling element to reinforce our *lack* of knowledge *in direct conjunction with the precision of our perception*, we are placed in a position of no longer searching for a kernel of truth. No longer is a layer of meaning being unveiled to show underneath some other fabrication given as unfabricated, some whole and cohesive analogue to the real or a metaphor for such. Instead—as if we had a choice, given the material at hand—we are placed, contradictorily, in the face of such confusion, or rather, of such lapses in the internal coherences of the narrative. We are disabled in the figuration of the plot. We are disabled in the figuring-out, the figuring-in: both are denied. Thus a fragment is given that is for once *not* a fragment of a whole that can be "spontaneously," "immediately," reconstituted in imagination. A fragment is given that does *not* control the meaning and stand for the meaning of the reconstituted whole. We are given fragmentation that does not define itself through mere difference from some apprehendable, understandable whole from which it happens to be disjoined. Therefore: the fragment is not given a cohesiveness and imaginary unity of materiality and meaning through a closure of its *place* in the whole of the narrative to identified *with* and identified *into*. Nor is the fragment given to stand in for, and represent, another, as if in the completeness of another's absence. Thus the two essential recuperations are made impossible tactics—impossible, period.

The nonidentificatory positions invoked by the mechanism of this television play, as described so far, rely on articulating a degree of convention inside certain dominant codes. To ignore those codes totally would simply place the drama in an arena of high style, in a stylistic outside of dominant codes that would, however, thereby reallow dramatic identification through fiction, plot, character. In the specific case of a form of expression, the stylistic, being outside the dominant codes (identification of, and into, the scene, the space), may be withheld for a time, after which the style is merely another code for understanding and, in fact, facilitates it. Thus, *The Cabinet of Dr. Caligari* is no more "difficult" that *Gertrud, The Man With a Movie Camera, Jaws, In the Realm of the Senses* or *Grease*.

On the other hand, the code of a kind of hypersubjectivity is necessitated precisely to make possible the complex distancing as I have been trying to outline it. For example, the sound becoming audible, as the head of the protagonist moves closer to the cassette, thus allows for the usual one-to-one identification into character, with the sound as "our sound"; the move from him to us, and us to him, is a device of convention *through which*, finally, an operation of unposition for the viewer/listener can be constructed/imposed. It would be an error, therefore, simply to distance the characters, the plot, the camera movements, and so on, through certain devices *prior* to such an imposition. For this reason—because of the fundamental misunderstanding that usually operates vis-à-vis distancing—we have had precious little Brechtian/Beckettian theater since the 1930s, in both the English- and the German-speaking countries: no materialist theater, no dialectical theater.

The man opens the window. *Point-of-view* shot of rain. And the sound of rain. But the fact that the rain fills the total frame (there is no window frame, no distant landscape, nothing to hold the perspective representation in place) places the shot of the rain into a particular position, one *not specific to the place outdoors for this particular scene*. It is, namely, the position of the rain, per se: not this rain, not this scene, not this signifier, or whatever rain would signify given the internal drama of a man in a room opening a window to find that outdoors it is raining, and that the rain has a sound which had not been heard indoors—indoors thus being, through that shot of the rain, established as a vacuum of sorts. Instead: rain, an unspecific, although materialist, term, and an unspecific, although materialist, event, that is materialist precisely through its not being constructed unproblematically within a drama. Instead another space, another event, with*out*. In that sense, the rain close-shot does not serve to naturalize the dramatization further, and to bring it forward; it serves, instead, to denaturalize the space, breaking into the continuance of dramatic flow or tension.

Similarly, an extreme close-up from above the cassette showing a "small gray rectangle on larger rectangle on seat"—abstracted so that "seat is a rectangle, flat and Lissitzky-like—denaturalizes the supposed function, within the dramatic narrative, of the cassette recorder. By

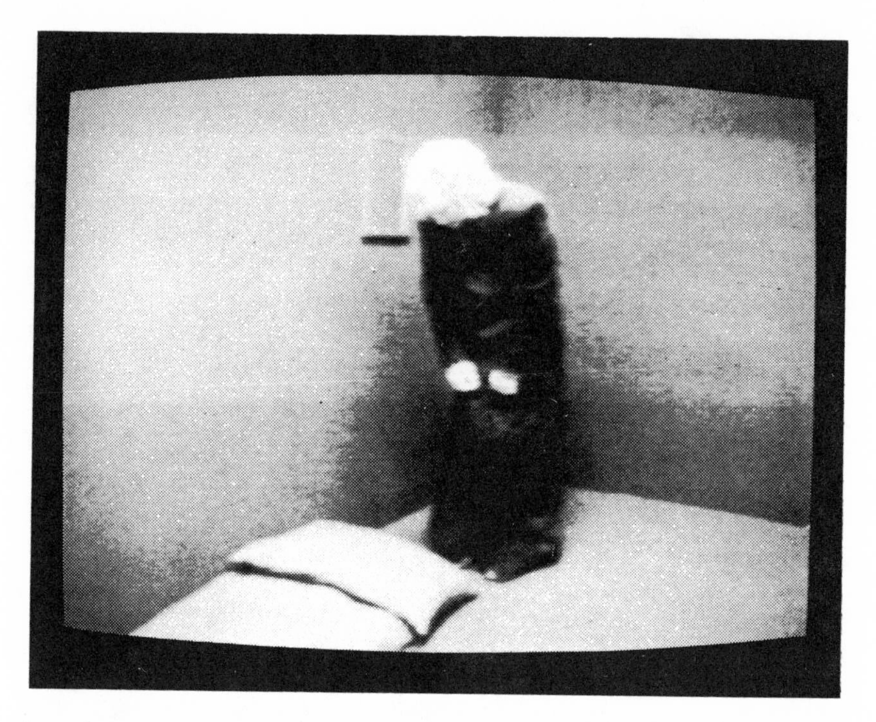

formalizing it, within this specific context of narrative drama, it becomes an other, outside, not befitting the space given for the story. The flattening of Renaissance perspective in the view perpendicularly from above breaks the possible identification with an anthropomorphic camera as metaphor or deputy for the consuming/consumed/unconsummated viewer.

A similar device is articulated later on: the Figure's point-of-view of the bed is given, then an extreme close-up of the mirror, then a medium of the Figure, pallet, mirror; the figure turns to the mirror. But the mirror close-up was pre-point-of-view, given in extreme close-up, as was the cassette recorder *before* the figure could (inside the dramatic narrative) have been placed to see realistically (into) the mirror. Thus the construction is foregrounded by being made difficult in viewing; the placement of a shot is not dictated, yet again, by the "psychological" necessities of character, drama, psychology, flow, rhythm, melody, form. "Cut to close-up of mirror reflecting nothing. Small gray rectangle, same dimensions as cassette, against larger rectangle of wall." Then, later, switching back to naturalism, a shot of the figure looking at the mirror, *from* the mirror.

Another *Geste* in *Ghost Trio*: the man walks, somewhat hunched, Nosferatu-like ("old man" posture) to his seat. During most of what has so

far been described, the narrator, or Voice, has been absent. The Figure takes the cassette; he hears the sound, and we hear the sound, of the Beethoven, which stops on a cut as he jerks right; he listens, and we hear footsteps. Shot 30: "F gets up, goes to door, opens it as before, looks out, stoops forward. Crescendo creak of door opening. Near shot of stool, cassette. F holding door open, stooping forward." Shot 31: "Cut to near shot of small boy full length in corridor before open door. Dressed in black oilskin with hood glistening with rain. White face raised to invisible F (5 seconds). Boy shakes head faintly. Face still raised (5 seconds). Boy turns and goes. Sound of receding steps, register from same position his slow recession till he vanishes in dark at end of corridor. 5 seconds on empty corridor."

Back to longshot of the whole scene. Sound comes on, after a time, camera still all the way out viewing the scene. Music louder of its "own" accord; no closing-in of the camera to get "nearer" to the "actual" cassette sound, nor a parallel soundtrack with its own logic throughout as something separate from the visual. Equally, it is not a matter of F's head moving closer to the cassette to hear. The volume increases over the scene, *then* the camera starts to move in, in contradistinction to the way sound operated for the hearer—us—via the camera movement or the protagonist, earlier on. As the camera moves in, the sound stays level for some moments, thus totally separated from the movement. "Hold till end of Largo." Then sound off. But even that hold on the sound, with a hold on the camera movement, does not persist. The sound, in its final hold, gets even louder, and one cannot register whether it is the sound of the video, or the music itself, as played by whomever. This confusion between the recorded movement by Beethoven, and the audio control of the video machine at the recording stage, disallows an easy separation of presentation and representation. The actual musical text, as given off, is positioned so that the difficulty of distinguishing difference persists—the difference "between" perception (what one hears) and knowledge (how that hearing was, what it was, in its construction or process). Thus there is also a resistance to the emptive, to the identificatory use of musical drama. Identification is contradicted and contraindicated by the machined operation of volume control, a break, again, in a possible naturalization of the place *of* the (musical) drama and the place *for* the viewer of that drama—hence, a break in the unquestioned flow of meaning and feeling. The construction, produced as a constant antithesis, or, better, nonthesis: the nonthetic, nonmimetic resistance.

Beckett, the producer said, made some changes from the printed text, in terms of the length of shots; but, when he did, he did that to every shot, thus not affecting the structure.

My first reading/viewing of initial moments of *Ghost Trio*, in 1976, had the protagonist moving the sound control on the cassette recorder up (louder), and yet the observer did not hear anything; sound seemed to "come in late," based on bad synchronization. The producer said it seemed all right to Beckett. It seems more than obvious now that is an artificiality within the scene, that the time-lapse operates solely as something preventing narrative closure and cohesion, as out-of-sync sound is used, similarly, in certain experimental films.

It must, finally, be mentioned that, although this work was specifically written for television, and constitutes, to my mind, the most radical use of television as a materialist practice (together with ". . . but the clouds . . .," which was shown in the same program of three plays for television by Samuel Beckett), it was, in fact, filmed in 16mm, and then video-recorded. The design and structure of *Ghost Trio*, a box for "the box," remained unimpaired by this use of film—whose greater depth, tonality and resolution befitted the precision of Beckett's written instructions—while retransmission through video befitted the conception and structure and view (point) of the work. Tryst indeed.

Notes

[1] John Calder (Beckett's publisher), in *The Journal of Beckett Studies*, No. 1, 1976, p. 119.

[2] Marthe Robert, *From Oedipus to Moses: Freud's Jewish Identity*, London, 1977, p. 187.

[3] Louis Althusser, Essays in *Self Criticism*, London, 1976, p. 85.

[4] Samuel Beckett, Letter to Peter Gidal, 12 September 1972.

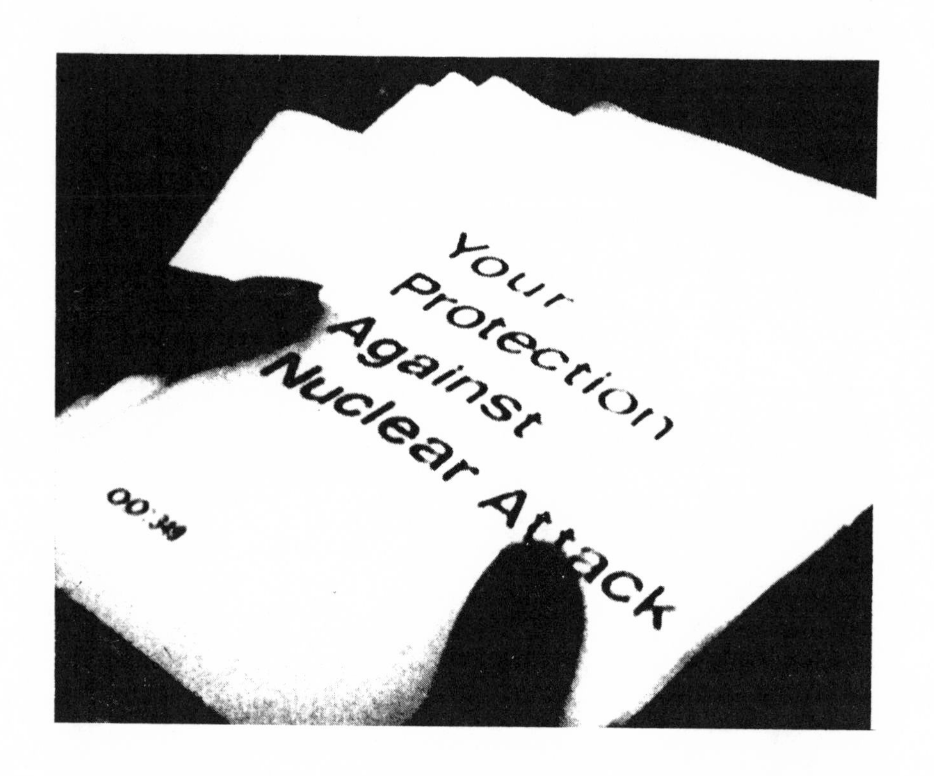
Your
Protection
Against
Nuclear Attack
00:39

NUCLEAR CONSCIOUSNESS ON TELEVISION

James M. Welsh

THERE CAN BE LITTLE DOUBT that the current decade is dominated by nuclear anxiety. On November 20th, 1983, a landmark day for ABC Television, the *New York Times* announced "Washington-Moscow Chill is Causing New Global Jitters," and Hedrick Smith's story discussed the placement of armed Cruise missiles in Britain and the planned deployment of Pershing 2 missiles in West Germany the following month to counteract the placement of Soviet SS-20 missiles in Eastern Europe.

It is hardly surprising that this high level of nuclear tension should be reflected by the entertainment industry and that there should be an interest on television in special programs treating nuclear disaster. In the same way such films as *On the Beach, Fail-Safe, Dr. Strangelove,* and *The War Game* reflected the nuclear tensions of the 1960's, *Special Bulletin, Testament,* and *The Day After* (broadcast on November 20th, 1983) reflected the "new global jitters" of the 1980's on television. Except for *The War Game,* which was originally intended for British television, the nuclear films of the 1960's though daring in content and message, are all clearly identifiable as commercial motion pictures, distinguished by identifiable stars, that could be and were marketed as entertainment films.

The television programs of the 1980's are rather more difficult to classify in such conventional terms. *Testament,* for example, was originally intended for "American Playhouse" on public television but was first released as a theatrical film in 1983. Almost from the beginning it was intended to serve both markets: "American Playhouse" agreed to cover two-thirds of the $2 million budget, but the rest of the production funding was covered by a British-based company, Entertainment Events, Ltd. The arrangement stipulated that *Testament* would first be released as a theatrical film and that at least fifteen months later it would be scheduled to air on PBS. Released in the United States by Paramount Pictures late enough in 1983 to capitalize on the attention that *The Day After* was then getting, *Testament* went on to earn an Oscar nomination for Jane Alexander.

The generic confusion with regard to *Testament* resulted from a creative funding ploy by producer-director Lynne Littman, after the original plan for a one-hour "American Playhouse" program that was funded for $500,000 expanded into a 90-minute program that was estimated to cost in excess of $750,000. Contract problems subsequently resulted since Jane Alexander and William Devane had worked on the project for the Screen Actors' Guild's "P" scale, a special public television rate that was lower than the scales usually negotiated for either commercial television or for theatrical motion pictures.

The crossover potential from public television to "serious" theatrical film markets was favorable in the case of *Testament*. Although Rita Kempley proclaimed the film to be "A Nuclear Dud" in *The Washington Post* (November 9, 1983), Robin Little, the editor of *Films in Review*, considered it a better picture than *The Day After* (which got far more attention in the press nationwide), and her interest and approval was later to be confirmed by the Motion Picture Academy.

The Day After, on the other hand, was designed for American commercial television, but, even so, it was later released as a theatrical film in Europe. The impact of the film's special effects, made possible by a production budget in excess of $7 million, no doubt assisted this film's crossover potential, since the formulaic weakness of the first hour was quite overwhelmed by the endemic anguish and surprising bleakness of the last hour, after the film had turned into a quasi-documentary about post-nuclear survival, and the often conventional characters, who never quite came alive to begin with, began to glow in the atomic wasteland.

In general, however, both *Testament* and *The Day After* observed the narrative and theatrical conventions ordinarily followed by television productions. *The War Game*, which was the first nuclear disaster film to be made for a mass television audience in 1965, was far different in its overall design, even though it was imitated by Nicholas Meyer in *The Day After* as Meyer attempted to structure and dramatize the immediate effects of a nuclear strike.

The War Game was never intended to be a little family drama with professional actors representing characters designed for audience indentification and sentimental empathy. Rather, it was a documentary that worked to achieve newsreel authenticity and to "educate" the viewer in the matter of nuclear warfare. In November of 1983, reflecting on the controversy surrounding the airing of *The Day After*, Peter Watkins, the director of *The War Game*, remarked: "there are as many possible forms of making film as there are blades of grass. But the media "—that is to say, the Media Establishment, those who make production and programming decisions for the major netwoks—"still clings virtuslly to one form, the theatrical mode." Twenty years ago Watkins effectively demonstated an alternative "mode" in the *The War Game*.

This realistic documentary was—and arguably still is—the most effective attempt to visualize the possible consequences of nuclear war and to capture a sense of the human suffering that would surely follow. *The War Game* was brilliantly experimental as a documentary film, but, unfortunately, it was too far advanced for its time and for its intended television context.

The War Game is not an easy film to describe. It is executed in a documentary, *cinema-verite* style (complete with interviews, some authentic, some scripted), and it is shot in "realistic" black-and-white stock. It therefore has the appearance of a newsreel, and the commentary is delivered in newsreel fashion by experienced BBC news readers. Though it follows a general chronological development, it eschews any kind of conventional or theatrical story-line. There certainly is broad-scale human drama but no banal melodrama here, and the delineation of conventional individual characters is totally dismissed. Watkins wants to confront the viewer with recognizable people, not theatrical characters. This is an informative film, in many ways a "factual" one, yet its plot device is borrowed from science fiction, dramatizing and evaluating events in Britain in a hypothetical future time-frame.

None the less, the speculation and warning Watkins offers is solidly based on facts (taken, for example, from Samuel Glasstone's *The Effects of Nuclear Weapons*, published by the U.S. Atomic Energy Commission) and carefully researched scientific calculation. At the same time, however, this is a film of conviction and moral outrage, though its tone appears to be well-reasoned, "scientific," objective, and controlled. Throughout it has the "feel" of a documentary television program examining a current event, yet it does not provide what Watkins has scornfully called "an intellectual wrap-up" at the end, and a list of six simplistic solutions for the problem it has introduced.

Although this film was eventually screened in commercial cinemas throughout Great Britain and earned an Academy Award as Best Documentary Feature for the year it was released, it was not originally intended for an audience of moviegoers. *The War Game* was produced by the British Broadcasting Corporation for a television audience, but when the BBC finally realized what sort of picture Watkins had made—a devastating indictment of nuclear brinksmanship and of Home Office policy that was obviously inadequate to deal with a crisis of the magnitude Watkins imagines in his film—the Director General of the BBC mandated a television ban on the film, a ban over which Watkins had absolutely no influence or control, a ban that is still in effect worldwide as of this writing, though the *Sunday Times* reported on 2 September 1984 that the BBC was willing to reverse the 19-year ban by the end of the year, pending final approval from the BBC Board of Governors.

Fortunately, however, the film was not simply shelved and forgotten. Eventually influential Members of Parliament were invited to see the film at private screenings, and many of them recognized its merits. Even those who opposed the film on political grounds could not deny its impact or effectiveness. After the ban was imposed by the BBC in 1965, a debate raged for months in the British press, keeping the issue of the film and its apparent censorship alive until finally a compromise was effected that enabled the British Film Institute to break precedent and enter into motion picture distribution. It was through the BFI that the film was therefore released. Otherwise, the strongest and most inventive film of Peter Watkins early career would not have been seen at all.

The complications surrounding this furor were first moral, then political. One of the film's first and most persistent opponents was Mrs. Mary Whitehouse, who had organized her "Clean-Up TV Campaign" to serve as a watchdog against violence on television. Other people were also particularly concerned about the effect the film might have on children. Similar questions were raised in 1983 during the weeks before *The Day After* was aired in the United States and helped to make the program a media "event," guaranteed to have impressive ratings.

Clearly, however, the Watkins film was halted more by political than by moral considerations. *The War Game* could be seen as an embarrassment to the British government. Of course, it was also said to be too graphic, too demoralizing, too violent, potentially disturbing to both the very young and to the old and infirm, "nervous" and "sensitive" people who, presumably,could not bear to watch such a grim spectacle.

Some dismissed the film as propaganda for the Campaign for Nuclear Disarmament, though Watkins was always careful to disassociate himself from that organization. The film obviously touched nerves, and Peter Watkins was ultimately frustrated by his own talent and by his ability to shock people into awareness. Ironically, the film simply made its point too well. That point was political, and Watkins was ahead of his time.

Part of the problem had to do with generic confusion. *The War Game* appears to be a documentary in its style and technique, but its point of view was clearly subjective. The filmmaker was obviously committed to a cause—the need for world peace and the intelligent containment of the arms race that fed and continues to feed the proliferation of nuclear weapons. *The War Game* was a subjective and avant-garde documentary, both daring and experimental in the way it extended the boundaries of the documentary film.

Although *The War Game* was produced as an educational film within the documentary division of the BBC, it was too clever, too experimental, and, finally, too effective. The objective approach one would ordinarily expect from an educational film was made to work a subjective response of revulsion. Watkins implanted much educational "data" and statistical information in his film, but this "objective" data is made to serve a rhetorical purpose through the use of devastating irony. The objective presentation of the data is purposefully contrasted with a subjective spectacle of suffering that is heart-wrenching. To this is added a structure of interviews that demonstrates an appalling level of public ignorance about nuclear destruction and its consequences.

Some fifteen years later the BBC produced its own version of *The War Game* following a more conventional formula, Mick Jackson's *Guide to Armageddon*, which examines what might happen if a one-megaton nuclear device were airburst above the dome of St. Paul's Cathedral, a central London landmark. As one might expect, the approach is properly "objective," and the speculated destruction is dryly (though not ironically) described. Pumpkins are shown being torn apart by shards of glass driven by the force of a nuclear windstorm, and viewers are invited to suppose that what happens to these pumpkins might also happen to human heads. This greengrocer approach to nuclear devastation is academic and detached, and not likely to inspire panic. This film is educational, but it fails to do what Peter Watkins did so brilliantly—to factor in human suffering.

In 1983 *The War Game* was also imitated by Nicholas Meyer in his film *The Day After*, which was made for the American Broadcasting Corporation, originally scheduled early in the year in one format, but only aired in an abbreviated form several months later. A comparison between *The Day After* and *The War Game* is interesting because of what it suggests properly belongs to commercial television and what does not. *The Day After,* comfortably set in the American midlands of Kansas and Missouri, squanders half of its time establishing sets of generally likeable characters who will later become the victims of nuclear war. The first half of the program appears to be a very conventional family drama, with adolescents in love and complications that would not be out of place in an evening soap-opera. The film works rather too hard to establish the myth of American family values.

The problem with this approach is that it fogs the issue, which can be humanized without the conventional television apparata, as Watkins demonstrated years ago. Obviously ABC wanted to present a product that would at first be recognizable to prime-time viewers and that product was then "sold" with the typical commercials that dominate American television. The network made a significant concession to taste, however, by refusing to sell commercial time that would interrupt the last half of the program, and at a midpoint the commercial product was transformed into an experimental and educational one. But not entirely, as we see a noble farm family trying to cope with the horrors of radiation poisoning and pillaging passers-by.

The Day After was successful in two respects: its graphic description of the nuclear strike against Kansas City and Minuteman silos nearby—the "*Vier Minuten Inferno*," as *Der Spiegel* described it—and in the television audience it captured, estimated to be as large as 100 million viewers for the initial American telecast. Thus, in one evening and at one single stroke, Nicholas Meyer, who admits to having been influenced by *The War Game*, managed to reach a larger audience than Peter Watkins has probably reached during his entire career as a filmmaker. Consequently, for a few hours the night of November 20th, 1983, the American public was sensitized as never before to the dangers of nuclear war.

Was *The Day After* truly "educational"? It surely contained important information, but one looks in vain for any significant impact the film might have had on American politics and foreign policy. Nine months later Ronald Reagan, whose stance towards the Soviet Union the world has judged to be antagonistic at best and whose commitment to the arms race seems clear, appeared to be no less popular than he had been before *The Day After* was shown. The film might have been educational, but it seems not to have influenced American voters to any significant, measurable degree. On December 5, 1983, *Time* magazine reported that a poll taken in Cambridge, Massachusetts, indicated that the film did not cause a significant number of people to change their opinions about nuclear defense policy. Fifty-eight percent of those polled continued to support American defense policies and forty-one percent still thought the United States was doing its best to avoid a nuclear confrontation. (Meyer's film left the question of who started the nuclear war purposefully ambiguous.) After the film five percent of those polled thought they could survive a nuclear war, down from seven percent before the film was aired. Perhaps, therefore, *The Day After* was ultimately regarded as merely dispensable entertainment, like almost all of what is usually seen on American television during prime-time. The audience is simply not conditioned to take such programming seriously.

The irony is that while Nicholas Meyer was preparing to shoot *The Day After* in Kansas, Peter Watkins was also at work on another nuclear film in London, a project that was destined to collapse before the first scheduled shooting date in July of 1982. This was not the first time Watkins had attempted to remake and update the message of *The War Game*. In 1968 he was approached by Sudwestfunk, a major West German television producer in Baden-Baden, but the project collapsed before research got underway. The nuclear issue was always at the forefront of the director's consciousness, however, and resulted in other related approaches as well. Norddeutscher Rundfunk, a television operation centered in Hamburg, first encouraged Watkins to pursue a film project that would "depict the possible consequences of a meltdown at the Indian Point reactor north of New York City," a proposal well ahead of *The China Syndrome* and the near meltdown at Three Mile Island. Norddeutscher Rundfunk cancelled this project, however, because, Watkins has written, "their technical expert had informed them that such an accident was not possible."

Two projects completed during 1975-76 were related. *Evening Land*, produced by the Danish Film Institute in 1976 and set in Copenhagen, was primarily concerned with radical terrorists, but the film was set politically against a strike of shipyard workers who refuse to build a nuclear submarine for the French navy. *The Trap*, a videodrama produced by Sveriges Radio in Sweden in 1975, is set in a bleak future world dominated by nuclear technology run amok, resulting in a police state intended to control problems of nuclear waste. *The Trap*, which aired to European markets, is remarkably innovative in its style, structure, and technique, but for that reason it is much too experimental for more conventional American markets. In the United States this Watkins work has only been seen in a few academic settings.

In 1978 Watkins negotiated with the Canadian Broadcasting Corporation to make another meltdown picture, this one keyed to the Luzon Peninsular reactor in the Philippines. Ironically, CBC cancelled this project two days before the Three Mile Island crisis outside of Harrisburg, Pennsylvania. In short, Watkins has most often met with frustration in his continued efforts to make nuclear-related films.

Then, late in 1981, Watkins was approached in Stockholm, where he had been working with the Swedish Film Institute, by Frank Allaun, a Labour Member of Parliament, who offered to establish a Peace Film Fund in Britain to finance an updated version of *The War Game*. Watkins was interested in this project and put together a research team in London during the spring of 1982. An agreement was reached with Central Television, a London producing firm, which offered to cover behind-the-camera expenses.

Developing his script for this film, Watkins settled upon a national approach to the problem, unlike the original *War Game*, which was centered in Kent. He also returned to an original concept that had not been fully developed in *The War Game* as completed, which was to make an individual family the focal point of the film. As the project developed and expanded in the director's mind, Central Television began to question production costs, since Watkins intended to shoot in as many as twenty-five locations in England and Scotland. The initial commitment called for a budget of 100,000 pounds, but Central claimed the expanded plan would extend that commitment to more than 750,000 pounds, which, the company claimed, would make production costs prohibitive.

This funding problem was not openly discussed with Watkins and his research crew until late May of 1982, when Central proposed bringing in a producer for the project. In June the project collapsed, only a few weeks before shooting was to have started. Central's expressed rationale for the cancellation was financial, though the expanded costs, even by their estimation, hardly seems immense.

The director's immediate response to this frustrating experience was to issue an angry press statement that was relatively ignored in Britain and only printed in its entirety in 1983 in *Literature/Film Quarterly*, an academic American film journal. Watkins was absolutely committed to the project and resolved to make the film in such a way that he himself would be utterly in control of production and distribution. To many, no doubt, his solution to the problem would seem idealistic and untenable, but by the summer of 1984 Watkins had scheduled the first shoot for his Nuclear War Film for September, after two years of an exhausting and global fund-raising effort that is, as far as I know, unique to the annals of filmmaking.

Just as, early in 1982, Watkins chose to expand the scope of his film to national proportions, when the project collapsed at mid-year, he began thinking in international terms. After he completed the film *Privilege* for Universal Pictures in London in 1966, Watkins became a man of the world. His next film, *The Gladiators*, he made in Sweden in 1968; *Punishment Park* was shot in California in 1970; *The Seventies People* was a documentary on teenage suicide made for Danish television in 1974; his biographical documentary *Edvard Munch*, begun in 1973 and completed in 1975, was made in Sweden and Norway. Meanwhile, the director's family had settled in France. Numerous lecture tours had put Watkins in touch with anti-nuclear groups throughout the United States, Canada, and Australia. The director's cosmopolitan orientation proved useful in his attempt to finance independently his Nuclear War Film.

Just as a network of individual peace groups had been set up in Great Britain to raise funds for the Central Television project in 1982, Watkins began networking a system of peace groups in eleven different countries worldwide with the idea of raising production money locally in twelve separate locations—two in the United States (Utica, New York, and Portland, Oregon, were the first choices), Scotland (where Watkins had already made contacts in 1982), France, Sweden, Germany, Kenya, Japan, Australia, Tahiti, Mexico, and one Soviet bloc country. (Watkins applied for and apparently was granted permission to film in the Soviet Union; contingency plans in case negotiations failed with the Soviets called for a back-up shoot in Czechoslovakia.)

Scott MacDonald, who helped to organize the fund-raising effort in Utica, described the film's approach as follows: "In all instances, Watkins will be working with families and family groupings who will speak for themselves about the multi-faceted international complexities of the arms race, about its real and potential human costs, about the ways in which conventionalized film and television language structures function to neutralize people's willingness to face the problems the arms race causes, and about ways in which average people can work constructively to disentangle themselves from the present, increasingly dangerous situation." This scheme therefore goes beyond the main purpose of *The Day After* and the original *War Game*, which was merely to alert viewers to the problem, without suggesting any particular means of dealing with it.

By late 1983 the initial fund-raising effort was well underway. On the West Coast organizers in Portland and Salem, Oregon and Seattle, Washington, had put together benefit dinners, screenings of *Culloden* and *The War Game*, and charity performances. Washington and Oregon chapters of the Physicians for Social Responsibility, the Federation of Reconciliation, the World Without War Council, and various church organizations were contacted. In Canada benefit screenings were organized in Montreal, Toronto, and Halifax, Nova Scotia. The National Film Board of Canada agreed to provide facilities, equipment, and film processing, and as of August 1984 Watkins had arranged to edit the film himself at the offices of the Film Board, probably in February of 1985. The Mexican Film Institute offered cameras for the Mexican shoot.

Artists also offered cooperation. The French theatre-director, filmmaker, and dramatist Armand Gatti volunteered his cultural centre in Toulouse (Atelier de Creation Populaire) and creative support, for example. In Oslo the Norwegian Artists for Peace organized a benefit performance and charity auction that raised 22,000 crowns. Among featured celebrities were singer Kari Svendsen and Swedish actress Bibi Andersson. For the New York group Dustin Hoffman offered a benefit performance of *Death of a Salesman*.

Such details represent only a fraction of the world-wide fund-raising effort. In other respects production costs will be kept to a bare minimum. In order to make the film Watkins will depend on much work that is volunteered, and, as is his preference, he will work with ordinary people rather than with trained actors—a method that has yielded extraordinary results for him in the past. Interviewed in November of 1983, Watkins pointed out that even though *The Day After* required a $7 million budget, he believed his film could be made for $500,000. By September of 1984, three weeks before the shooting was to begin in upstate New York, approximately $35,000 had been raised for the Utica segment, enough to cover the family footage and evacuation footage as well. Adjustments will be made at the other shooting sites according to production funds raised by the individual peace groups.

Although certainly influenced by the work Peter Watkins had done some twenty years earlier, *The Day After* was a worthy film. Although compromised by its made-for-television design, it was a timely and significant picture, certainly superior if measured against those products usually made for American television markets. And yet, as television critic Jay Cocks and others have asserted, it is not so effective as *The War Game*, which it imitates. It did not take into consideration the long-range problems that Watkins attempted to document—the problem of disposing of thousands of corpses and identifying the dead, the shortage of pain-killing drugs, the utter breakdown of civil authority, the food riots, the collapse of civilized authority, the food riots, the collapse of civilized human behavior, the bleakness of the interviews with children at the end of *The War Game* who tell the viewer "I don't want to be nothing" and "I don't want to grow up." Perhaps some of that despair was captured, however, by the church service in *The Day After* conducted by a minister whose faith in a just and loving God has obviously been shaken, by the release of Steven Klein and Denise Dahlberg from the hospital facilities in Lawrence, who have both been dangerously radiated and are presumably going home to die, and by the final embrace between Jason Robards and the survivor he encounters in the rubble of his "home" in Kansas City.

Even so, *The Day After* ultimately did not advance the accomplishment of *The War Game*, as the latest Watkins efforts may do. It merely brought a similar message to a larger audience that was sensitized and perhaps "numbed" momentarily. Its focal point was local. The focal point of *The Nuclear War Film* that Peter Watkins has in progress is global. This time Watkins will not concentrate on the immediate effects of blast and firestorm on a local community. Rather, he will concentrate on the feelings and beliefs of ordinary people about the nuclear dilemma, and he will attempt to show what ordinary people can do to influence their goverments in an effort to avoid a nuclear disaster in a film that he hopes to have completed by August 6th, 1985, to mark the twenty-year anniversary of the Hiroshima bombing.

Financial support appears to be reasonably strong in most of the selected locations. What is particularly advantageous about this funding scheme is that the project could go on, even if there were financial shortfalls in some of the proposed locations. In other words, the film is likely to be completed. Certainly it will not be destroyed by administration fiat. I know of no other comparable grassroots funding scheme for any motion picture made by an Academy Award-winning director.*The Nuclear War Film* is therefore unique, and without precedent. It may never reach 100 million viewers on any given evening, but its fate will be entirely in the hands of the director.

THE CASE OF THE A-BOMB FOOTAGE

Erik Barnouw

IT WAS IN 1970, a quarter of a century after the footage was shot, that the documentary film *Hiroshima-Nagasaki, August 1945* had its premiere and won an audience, an international one, as it turned out. In recounting the origin and history of this film I want to emphasize the extraordinary twenty-five year hiatus. It seems to me to have implications for filmmakers and film scholars and perhaps for the democratic process.[1]

I became involved in this story in its later stages, as producer of the 1970 compilation documentary, and this involvement came about almost by accident. Before I explain how this happened, let me go back to the beginning of the story, as I have been able to piece it together over the years.

In August 1945, after the two atom bombs had been dropped on Hiroshima and Nagasaki, a Japanese film unit named Nippon Eiga Sha was commissioned by its government to make a film record of the effects of the devastating new weapon. Nippon Eiga Sha was a wartime amalgamation of the several newsreel and documentary units that had existed before the war. They had been nationalized for war purposes.

The man entrusted with the making of the film was Akira Iwasaki, a film critic, historian, and occasional producer. The choice of Iwasaki for the assignment was significant. During the 1930s he had been the leader of a leftist film group called Prokino or Proletarian Film League, similar to the Workers Film and Photo Leagues in the United States. Being antimilitarist, Prokino had been outlawed shortly before the war, and some of its members had been jailed under a preventive-detention law. Iwasaki himself had spent part of the war in prison. The fact that he had regained standing and was given the film assignment reflected the turbulent situation in the final days of the war, and the extent to which the military had already lost status.[2]

Because of the breakdown of transport and the difficulty of obtaining adequate supplies, it took the Nippon Eiga Sha film crews some time to reach their locations, but they were at work in Hiroshima and Nagasaki when the American occupation forces arrived. What happened then has been described by Iwasaki. "In the middle of the shooting one of my cameramen was arrested in Nagasaki by American military police. . . I was summoned to the GHQ and told to discontinue the shooting." The filming was halted, but Iwasaki says he remonstrated, and 'made arguments' with the occupation authorities. "Then" he writes,

came the group of the Strategic Bombing Survey from Washington and they wanted to have a film of Hiroshima and Nagasaki. Therefore the U.S. Army wanted to utilize my film for the purpose, and changed its mind. Now they allowed me or better ordered me to continue and complete the film.[3]

During the following weeks, under close US control, much additional footage was shot. All was in black-and-white; there was no colour film in Japan at this time. As the shooting progressed, the material was edited into sequences under the overall title 'Effects of the Atomic Bomb'. There were sequences showing effects on concrete, effects on wood, effects on vegetation, and so on. The emphasis was on detailed scientific observation. Effects on human beings were included, but sparsely. Survivors on the outer fringes of the havoc were photographed in improvised treatment centres, but here too the guiding supervisory principle was scientific data-gathering rather than human interest. The interests of the camera teams were to some extent at variance with this.

The edited material had reached a length of somewhat less than three hours when the saga entered a new phase. Occupation authorities suddenly took possession of the film—negative, positive, and out-takes, and shipped it to Washington. Film and all related documents were classified SECRET and locked away, disappearing from view for almost a quarter of a century. Most people, including those in the film world, remained unaware of its existence. Apparently a few feet were released for Army-approved uses, and the project was briefly mentioned in *Films Beget Films,* by Jay Leyda, published in 1964, a book that began as a memorandum for the Chinese government of the values of film archives.[4] But whether the earliest Hiroshima and Nagasaki footage still existed was an American military secret. With later colour footage of the ruins making an appearance and to some extent satisfying curiosity, the missing footage did not become an issue in the United States.

Until 1968 I was oblivious to its existence. But early that year a friend, Mrs. Lucy Lemann, sent me a newspaper clipping she had received from Japan, which excited my interest. It was from the English-language *Asahi Evening News,* and reported that the footage shot in Hiroshima and Nagasaki in 1945 by Japanese cameramen had been returned to Japan from the United States and that the government would arrange a television screening "after certain scenes showing victims' disfiguring burns are deleted". The item also stated that the film would later be made available on loan to "research institutions" but it added: "In order to avoid the film being utilized for political purposes, applications for loan of the film from labor unions and political organizations will be turned down"[5].

I was at this time chairman of the film, radio and television division of the Columbia University School of the Arts, and had organized a related unit called the Center for Mass Communication, division of Columbia University Press producing and distributing documentary films and recordings. Naturally the clipping seemed to demand some investigation or action. Mrs. Lemann was a contributor to the World Law Fund, and at her suggestion I wrote for further information to Professor Yoshikazu Sakamoto, Professor of International Politics at the University of Tokyo, an associate of the Fund. His prompt reply said that the Japanese had negotiated with the US Department of State for the return of the film but the Department of Defense was thought to control it. The material sent to Japan was not the original nitrate film but a safety-film copy.

Somewhat impulsively, I wrote a letter on Columbia University stationery, signed as 'Chairman, Film, Radio, Television', addressed to 'The Honorable Clark M. Clifford, Secretary of Defense', with the notations that 'cc' should go to the Secretary of State Dean Rusk and to Dr. Grayson Kirk, President of Columbia University. The letter asked whether Columbia's Center for Mass Communication might have the privilege of releasing in the United States the material recently made available for showing in Japan.[6] I felt a bit flamboyant in this, but sensed I had nothing to lose. I scarcely expected results. But to my amazement, a letter arrived within days from Daniel Z. Henkin, Deputy Assistant Secretary of Defense, stating that the Department of Defense had turned the material over to the National Archives and that we could have access to it there.[7] So it was that early in April 1968 I found myself with a few associates in the auditorium of the National Archives in Washington, looking at some two hours and forty minutes of Hiroshima and Nagasaki footage. We also examined voluminous shot lists in which the location of every shot was identified and its content summarized and indexed. Every sheet bore the classification stamp SECRET but this had been crossed out and another stamp substituted: 'NOT TO BE RELEASED WITHOUT APPROVAL OF THE D.O.D.' There was no indication of the date of this partial declassification. We guessed that some routine declassification time-table had taken effect, but without public announcement. Perhaps we were merely the first to have inquired about the material.[8]

Some in our group were dismayed by the marginal quality of much of the film, a result, perhaps, of the circumstances under which it had been shot and the fact that we were looking at material some generations away from the original. But this quality also seemed a mark of authenticity; and it seemed to me that enough of the footage was extraordinary in its power, unforgettable in its implications, and historic in its importance, to warrant our duplicating all of it. A grant from Mrs. Lemann to Columbia University Press made it possible to order a duplicate negative and work-print of the full two hours and forty minutes, along with photostats of the priceless shot lists. During the summer of 1968 all this material arrived at Columbia University from the National Archives, and we began incessant study and experimentation with the footage, with constant reference to the shot lists and other available background information.[9]

The footage contained ruins in grotesque formations, and endless shots of rubble. At first we were inclined to discard many of the less striking rubble sequences, but when we learned that one had been a school (where most of the children had died at their desks), and one had been a prison (where 140 prisoners had died in their cells), and another had been a trolley car (whose passengers had evaporated, leaving in the rubble a row of their skulls and bones), even the less dramatic shots acquired new meaning. Eventually a montage of such rubble shots, linked with statistics about the people annihilated or injured, and the distance of each location from the centre of the blast, became a key sequence in the film.

The paucity of what we called 'human-effects footage' troubled us deeply. We felt that we would have to cluster this limited material near the end of our film for maximum effect, but meanwhile we resolved on a sweeping search for additional 'human-effects footage.' We wrote to the Defense Department asking whether additional material of this sort had perhaps been held back. The Pentagon's staff historian answered, assuring us nothing was being held back, and adding: "Out-takes from the original production no longer exist, having probably been destroyed during the conversion from acetate (*sic*) to safety film—if they ever were turned over to the US Government at all".[10] This curious reply made us wonder whether footage such as we hoped to find might still exist in Japan or might be held by people in the United States who were in Japan during the Occupation. Barbara Van Dyke, who became associate producer for our film, began writing letters to a long list of people, asking for information on any additional footage. In the end this search proved fruitless; we found we had to proceed without additional 'human-effects footage.'

One of those to whom she wrote was the Japanese film critic and historian Akira Iwasaki, the original producer. His name was not mentioned in documents received from the Defense Department or from the National Archives, but was suggested by the writer Donald Richie, a leading authority on Japanese cinema, as a likely source of information.[11] Iwasaki did not reply to our inquiry; he explained later that he had doubted the 'sincerity' of our project.

Her search did produce one extraordinary find. One of the occupants of the observation plane that followed the Enola Gay, the bomb-dropping plane, to Hiroshima was Harold Agnew, who later became head of the Los Alamos laboratory. As a personal venture he had taken with him a 16 mm camera. The very brief sequence he brought back provides an unforgettable glimpse of the historic explosion, and the shuddering impact of the blast on the observation plane itself, which seems likely for a moment to be blown to perdition. From Mr. Agnew we acquired a copy of this short sequence.

Our first rough assembly was some forty minutes long, but we kept reducing it in quest of sharper impact. What finally emerged, after more than a year of experimentation, was a quiet 16 minute film with a factual, eloquently understated narration written by Paul Ronder, and spoken by him and Kazuko Oshima. Ronder and Geoffrey Bartz did the editing, with musical effects by Linea Johnson and Terrill Schukraft. We consulted at various stages with Albert W. Hilberg, MD, and historian Henry F. Graff. We were not sure the film would have the effect we hoped for, but our doubts were soon resolved.

After several small screenings we arranged a major preview at the Museum of Modern Art in February 1970, to which the press was invited. The auditorium was jammed, and at the end of the showing the audience sat in total silence for several seconds. We were at first unsure what this meant, but the comments soon made clear what it meant. Later that day the UPI ticker carried a highly favorable report that treated the film as a major news event, mentioning the address of the Center for Mass Communication and the print sale price, $96. Two days later cheques and orders began arriving in the mail and continued without promotional effort on our part, at the rate of a hundred a month. In five months almost five hundred prints were sold to film libraries, colleges, school systems, clubs, community groups and churches. Every screening seemed to bring a surge of letters and orders. Foreign sales came quickly.

Two things amazed us: (1) the electric effect on audiences everywhere, and (2) the massive silence of the American networks. All had been invited to the press preview; none had attended. Early in the morning after the resounding UPI dispatch, all three commercial networks phoned to ask for preview prints and sent motorcycle couriers to collect them, but this was followed by another silence. By making follow-up phone calls we learned that CBS and ABC were 'not interested.' Only NBC thought it might use the film, if it could find a 'news hook.' We dared not speculate what kind of event this might call for.

The networks' attitude was, of course, in line with a policy all three had pursued for over a decade, of not broadcasting documentaries other than their own. We at Columbia University were outraged at the network policy.[12] We had half expected that the historic nature of the material would in this case thrust the policy aside. But we were for the moment too busy filling non-television orders to consider any particular protest or action.

Then a curious chain of media phenomena changed the situation. On 5 April 1970, the Sunday supplement *Parade,* which generally gave its chief attention to the romantic aberrations of the mighty, carried a prominent item about *Hiroshima-Nagasaki, August 1945,* calling it unforgettable, and necessary viewing for the people of any nation possessing the bomb.[13] This apparently caused the editors of the Boston *Globe,* which carried *Parade,* to wonder why television was ignoring the film. They made phone calls to nuclear scientists and others, asking their opinions on the matter, and reached several who had attended our previews. The result was a lead editorial in the *Globe* headed: 'HIROSHIMA-NAGASAKI, AUGUST 1945—NOT FOR SENSITIVE U.S. EYES'. It quoted Dr. S. E. Luria, Sedgwick Professor of Biology at Massachusetts Institute of Technology, describing the film as 'a very remarkable document' and adding, "I wish every American could see it, and particularly, every Congressman." Norman Cousins described himself as 'deeply impressed.' The *Globe* ended its editorial with a blast at the networks for ignoring the film[14]. *Variety,* the show-business weekly, was interested in the *Globe's* 'needling' of the networks, and featured the issue in a special box in its next edition.[15] This brought sudden action from National Educational Television, which a few days later signed a contract to broadcast the film in early August, twenty-five years after the dropping of the bombs. No sooner had the contract been signed than NBC announced that it wanted the film for use on its monthly magazine series, *First Tuesday.* When Sumner Glimcher, manager of the Center for Mass Communication, explained that the film was committed to NET, he was asked if we could 'buy out' NET so that NBC could have the film, but we declined to try.

As the issue of a US telecast was moving to a resolution, we were aware of parallel, and apparently more feverish, developments in Japan. Our first inkling of what was happening in Japan came at the Museum of Modern Art preview, at which we were approached by a representative of TBS (Tokyo Broadcasting System) one of Japan's commercial systems, with an offer to purchase Japanese television rights. To be negotiating such a matter seemed strange, in view of the Japanese government's announced plans for a television screening, but the TBS man was persistent and eager, and we finally signed an agreement authorizing a telecast, with an option to repeat. The telecast took place 18 March 1970, and the option to repeat was promptly exercised. We gradually became aware, through bulletins from Japan, of the enormous impact made by these telecasts. The government-arranged showing had taken place earlier over NHK, the government network, but had included little except rubble. Human beings had been excised "in deference to the relatives of the victims", but this action had brought a storm of protest. It was against this background that TBS had negotiated for our film. It also gave our film, which made use of everything that the NHK telecast had eliminated, an added impact. Professor Sakamoto of the University of Tokyo began sending us voluminous translations of favourable reviews and articles, one of which paid special tribute to Columbia University for showing the Japanese people "what our own government tried to withhold from us." The reviews included major coverage in a picture magazine of the *Life* format. Viewing statistics were provided. The *Mainichi Shimbun* reported that the film "caused a sensation throughout the country", while in Hiroshima "the viewing rate soared to four times the normal rate".[16] The *Chugoku Shimbun* reported:

At the atomic injury hospital in Hiroshima last night, nine o'clock being curfew time, all was quiet. Only in one room on the second floor of the west wing, the television diffidently continued its program. . . They had obtained special permission from the doctors. . . The first scene was of ruins. 'That's the Aioi Bridge.' 'That's the Bengaku Dome.' The women follow the scenes. Even the Chinese woman who had not wanted to see is leaning from her bed and watching intently. . . The scene of victims which has elicited so much comment is now on. 'That's exactly how it was,' they nod to each other. However, when the film was over they contradicted their words and said, 'It was much, much worse.'[17]

A letter came from the Mayor of Hiroshima. The city would mark the twenty-fifth anniversary of the bomb with a major observance including a long television programme, and wanted to include material from our film.

The most gratifying response came from Akira Iwasaki, who after a lapse of almost twenty-five years had seen his footage on television. His role in the project was not credited, and he might have been expected to resent this, but no sign of resentment appeared. He wrote us a long letter expressing his gratitude and appreciation for how we had used the material. He also published a long review in a leading Japanese magazine, describing his reaction.

I was lost in thought for a long time, deeply moved by this film. . . I was the producer of the orginal long film which offered the basic material for this short film. That is, I knew every cut of it. . . yet I was speechless. . . It was not the kind of film the Japanese thought Americans would produce. The film is an appeal or warning from man to man for peaceful reflection—to prevent the use of the bomb ever again. I like the narration, in which the emotion is well controlled and the voice is never raised. . .

That made me cry. In this part, the producers are no longer Americans. Their feelings are completely identical to our feelings.[18]

The impact of the film was further illuminated by a bizarre incident. At my Columbia University office a delegation of three Japanese gentlemen was announced, and ushered in, all impeccably dressed. One member, introducing the leader, identified him as a member or former member of the Japanese parliament, representing the Socialists. The leader himself then explained that he came on behalf of an organization called the Japan Congress Against A and H Bombs, also known as Gensuikin. In this capacity, he had three requests to make. First, as a token of appreciation for what we had achieved with our film *Hiroshima-Nagasaki, August 1945,* would I accept a small brooch as a gift to my wife? Puzzled and curious, I accepted.

Second, would I consider an invitation to speak in Hiroshima on the twenty-fifth anniversary of the dropping of the bomb, in the course of the scheduled observances? I hesitated—the suggestion raised endless questions in my mind, but I said I would consider. The leader seemed reassured, and said I would receive a letter.

Then came the third request. Would he be permitted to purchase six prints of *Hiroshima-Nagasaki, August 1945?* I explained that we sold prints at $96, for non-profit use, making no discrimination among buyers; anyone could buy. With an audible sigh of relief, he suddenly unbuttoned his shirt, ripped out a money belt, and produced six pristine $100 bills. Accustomed to dealing with checks and money orders, it took the office a while to round up the $24 'change.' We handed him the six prints. One member of the delegation had a camera ready; photographs were taken and the group departed. But a few days later we received a letter from another organization with a very similar name—the Japan Council Against Atomic and Hydrogen Bombs, or Gensuikyo. It wished the right to translate our film into Japanese, without editing change. Again we wrote to Professor Sakamoto of the University of Tokyo for enlightenment. Again he responded promptly.

. . . the movement against atomic bombs has been split into two groups since early in the 1960's, the immediate cause being the difference in attitude toward the nuclear tests carried on by the Soviet Union. The Japan Congress Against A and H Bombs, which politically is close to the Social Democrats, is against all nuclear tests, regardless of nation. The Council Against Atomic and Hydrogen Bombs, the other body, is close to the Communist Party, and is opposed to nuclear tests by the United States, but considers tests by the Socialist countries undesirable but necessary. . . The Council is a somewhat larger organization than the other. Many efforts have been made in the past to merge the two bodies but none have been successful to date.[19]

In the following weeks we were bombarded by both Congress and Council with cabled requests about prints, translation rights, and 8mm rights. To our relief the issue was resolved, or apparently resolved, by another news item from Professor Sakamoto. He reported the revelation that in 1945 a Nippon Eiga Sha technician, fearing that the American military would seize and remove the footage, had secreted a duplicate set in a laboratory ceiling. He had now made this known.[20] From then on we referred Japanese inquirers to this 'newly available' resource. Apparently the Defense Department's suspicion, expressed in the letter from the Pentagon historian, had had some validity.

On August 3, 1970, *Hiroshima-Nagaski, August 1945* had its American television premiere over National Educational Television, giving the system one of its largest audiences to date. 'Hiroshima Film Gets Numbers,' *Variety* reported.[21] NBC's *Today* programme and CBS *Evening News With Walter Cronkite* had decided, at the last moment, to carry news items about the event, using short clips and crediting NET and Columbia University. NET's Tampa Bay outlet did a delayed telecast via tape, after deleting some of the 'human-effects footage'. So far as we could learn, all other stations carried the full film. The telecast won favourable reviews across the nation, acclaiming NET's decision to show it.

To my disappointment, NET coupled the film with a panel discussion on the subject, 'Should we haved dropped the bomb?' It was an issue I had resolutely kept out of the film, even though most members of our group wanted the film to condemn Truman's actions. I did not myself see how President Truman, in the situation existing at that time, could have refused a go-ahead. But this seemed to me irrelevant to our film, a bygone issue, already endlessly discussed. To centre on it now seemed to me an escape into the past. To me the Hiroshima-Nagasaki footage was meaningful for today and tomorrow, rather than for yesterday.

During the research for my books on the history of American broadcasting, especially *The Image Empire*, I became chillingly aware of how often in recent years men in high position have urged use of atomic weapons. French Foreign Minister Georges Bidault has said that Secretary of State John Foster Dulles, during the Dien Bien Phu crisis, twice offered him atom bombs to use against the beleaguering Vietnamese forces, but he demurred.[22] Oral histories on file at the Dulles Collection in Princeton make clear that Dulles made the offer on advice received from the Joint Chiefs of Staff. Apparently Bidault's refusal (not President Eisenhower's as some writers have assumed) averted another holocaust. During the Quemoy-Matsu confrontation, use of an atom bomb was again discussed.[23] In 1964, Barry Goldwater felt that use of a 'low-yield atomic device' to defoliate Vietnamese forests should be considered.[24] (He later emphasized that he had not actually recommended it.) More recently there has been widespread discussion of proposed world strategies based on 'tactical' nuclear weapons—a term meant to suggest a modest sort of holocaust, but actually designating bombs equivalent in destructive power to the Hiroshima bomb. A more advanced bomb now equals 2500 Hiroshima bombs, as our film makes clear. Victory with such weapons would apparently win an uninhabitable world.

Such proposals can only be made by people who have not fully realized what an atomic war can be. When I first saw the Hiroshima-Nagasaki footage, I became aware how little I had comprehended it. Yet this footage gives only the faintest glimpse of future possibilities.

Why, and by what right, was the footage declared SECRET? It contains no military information, the supposed basis for such a classification. Then why the suppression? Why the misuse of the classification device? The reason most probably was the fear that wide showing of such a film might make Congress less ready to appropriate billions of dollars for ever more destructive weapons.

I produced the short film *Hiroshima-Nagasaki, August 1945* with the hope that it would be seen by as many people as possible on all sides of every iron curtain. If a film can have the slightest deterrent effect, it may be needed now more than ever. Fortunately it is achieving a widening distribution.

Although I did not accept the invitation to the 1970 Hiroshima observances, I have visted Japan twice since then, had long talks with Akira Iwasaki, met one of the cameramen in his 1945 unit, and visited the generously helpful Professor Sakamoto. I continued to correspond with Iwasaki until his death on 16 September 1981.

Notes

[1] *Hiroshima-Nagasaki, August 1945* is now widely available. Primary distributor is the Museum of Modern Art, 11 West 53, New York, 10019, which offers prints for life-of-the-print lease or per-day rental. Per-day rentals are also offered by many film libraries—more than 1000 prints are in circulation. The film's history was discussed by the author in a public lecture in Philadelphia, 11 February 1980, at the Walnut Street Theater; the present report is based on that lecture.

[2] Interviews with Akira Iwasaki, Fumio Kamei, Ryuchi Kano, in Tokyo, February 1972; Iwasaki, 'The Occupied Screen', *Japan Quarterly*, 25, 3, July-September 1978.

[3] Letter, Iwasaki to Barbara Van Dyke, 15 March 1970, Hiroshima-Nagasaki file, Barnouw Papers. All letters, memoranda, and press excerpts quoted in the present report are available in a file of some 300 items deposited with the Barnouw Papers in Special Collections, Columbia University Library. Photocopies of the entire file have also been deposited at the Museum of Modern Art, primary distributor of the film, at the Motion Picture, Broadcasting and Recorded Sound Division of the Library of Congress, Washington and Imperial War Museum, London. The file will be referred to hereafter as 'HN file'.

[4] Leyda, Jay (1964) *Films Beget Films* (New York).

[5] *Asahi Evening News*, 26 January 1968.

[6] Letter 8 March 1968, HN file.

[7] Letter 19 March 1968, HN file.

[8] The lists were on large file cards, cross-referenced for such topics as 'Atomic: physical aspect', 'Shadow effects', 'Shrine', 'Debris', 'Civilians: Jap.'. Copies of 53 such cards are in HN file.

[9] The 2 hours, 40 minutes of *Effects of the Atomic Bomb* held in the National Archives were in the form of 35 mm acetate preservation material made from the original, which was on unstable nitrate and no longer existed. We worked from 16 mm material made from the National Archives holdings.

[10] Letter, R. A. Winnacker to author, 27 June 1968. The word 'acetate', used in error, should have been 'nitrate'. HN file.

[11] Donald Richie, co-author with Joseph L. Anderson of the 1959 *The Japanese Cinema* (Rutland & Tokyo) and long-time film critic in Tokyo, with occasional service as visiting film curator at the Museum of Modern Art, New York.

[12] For genesis of the policy see BARNOUW (1975) *Tube of Plenty*, pp. 269-270 (New York).

[13] *Parade*, 5 April 1970.

[14] *Boston Sunday Globe*, 5 April 1970.

[15] *Variety*, 22 April 1970.

[16] *Mainichi Shimbun*, 20 and 22 March 1970.

[17] *Chugoku Shimbun*, 19 March 1970.

[18] Iwasaki, 'It all started with a letter', *Asahi-Graph*, 3 April 1970.

[19] Letter, Professor Yoshikazu Sakamoto to author, 4 May 1970. HN file.

[20] Letter, Sakamoto to author, 8 June 1970. HN file.

[21] *Variety*, 5 August 1970.

[22] The Bidault statement is in Drummond, Roscoe and Gaston Coblentz (1960) *Duel at the Brink: John Foster Dulles' Command of American Power*, pp. 121-122. (Garden City, New York), Bidault repeated the statement in the Peter Davis documentary film *Hearts and Minds*.

[23] Statements by Nathan Twining and George V. Allen in Dulles Oral History Collection, Princeton University, re Dien Bein Phu and Quemoy-Matsu deliberations.

[24] The Goldwater statement was in a 1964 ABC-TV interview with Howard K. Smith and became a major focus of the 1964 presidential campaign won by Lyndon B. Johnson.

GUERRILLA TELEVISION

Deirdre Boyle

VIDEO PIONEERS didn't use covered wagons; they built media vans for their cross-country journeys colonizing the vast wasteland of American television. It was the late Sixties, and Sony's introduction of the half-inch video Portapak in the United States was like a media version of the Land Grant Act, inspiring a heterogeneous mass of American hippies, avant-garde artists, student-intellectuals, lost souls, budding feminists, militant blacks, flower children, and jaded journalists to take to the streets if not the road, Portapak in hand, to stake out the new territory of alternative television.

In those early days everyone with a Portapak was called a "video artist." Practioners of the new medium moved freely within the worlds of conceptual, performance, and imagist art as well as documentary. Skip Sweeney of Video Free America, once called the "King of Video Feedback," also designed video environments for avant-garde theater (*AC/DC*, *Kaddish*) and collaborated with Arthur Ginsberg on a fascinating multi-monitor documentary portrait of the lives of a porn queen and her bisexual, drug addict husband, *The Continuing Story of Carel and Ferd*. Although some artists arrived at video having already established reputations in painting, sculpture or music, many video pioneers came with no formal art training, attracted to the medium because it had neither history nor hierarchy nor strictures, because one was free to try anything and everything, whether it was interviewing a street bum (one of the first such tapes was made by artist Les Levine in 1965) or exploring the infinite variety of a feedback image. Gradually two camps divided: the video artists and the video documentarists. The reasons for this fissure were complex, involving the competition for funding and exhibition, a changing political and cultural climate, and a certain disdain for nonfiction work as less creative than "art"—an attitude also found in the worlds of film, photography and literature. But in video's early years, guerrilla television embraced art as documentary and stressed innovation, alternative approaches, and a critical relationship to television

Just as the invention of movable type in the 15th century made books portable and private, video did the same for the televised image; and just as the development of offset printing launched the alternative press movement in the Sixties, video's advent launched an alternative television movement in the Seventies. Guerrilla television was actually part of that larger alternative media tide which swept over the country during the Sixties, affecting radio, newspapers, magazines, publishing, as well as the fine and performing arts. Molded by the insights of Marshall McLuhan, Buckminster Fuller, Norbert Wiener, and Teilhard de Chardin; influenced by the style of New Journalism forged by Tom Wolfe and Hunter Thompson; and inspired by the content of the agonizing issues of the day; video guerrillas set out to tell it like it is—not from the lofty, "objective" viewpoint of TV cameras poised to survey an event, but from within the crowd, subjective, and involved.

Video Gangs

For baby boomers who had grown up on TV, having the tools to make your own was heady stuff. Most early video makers banded together into media groups; it was an era for collective action and communal living, where pooling equipment, energy, and ideas made more than good sense. But for kids raised on "The Mickey Mouse Club," belonging to a media gang also conferred membership in an extended family that unconsciously imitated the television models of their youth. Some admitted they were attracted by the imagined "outlaw" status of belonging to a video collective, less dangerous than being a member of the Dalton gang—or the Weather Underground—and probably more glamorous. As video collectives sprouted up all over the country, the media gave them considerable play in magazines like *Time, Newsweek, TV Guide, New York* and *The New Yorker*, to name a few. They celebrated the exploits of the video pioneers in mythic terms curiously reminiscent of the opening narrations of TV Westerns. Here's an example from a 1970 *Newsweek* article:

Television in the U.S. often resembles a drowsy giant, sluggishly repeating itself in both form and content season after season. But out on TV's fringe, where the viewers thus far are few, a group of bold experimenters are engaged in nothing less than an attempt to transform the medium. During the past few years, television has devleoped a significant avant-garde, a pioneering corps to match the press's underground, the cinema's verite, *the theater's off-off-Broadway. Though its members are still largely unknown, they are active creating imaginative new programs and TV "environments"—not for prime time, but for educational stations, closed-circuit systems in remote lofts and art galleries and, with fingers crossed, even for the major networks.*

Video represented a new frontier—a chance to create an alternative to what many considered the slickly civilized, commercially corrupt, and aesthetically bankrupt world of Television. Video offered the dream of creating something new, of staking out a claim to a virgin territory where no one could tell you what to do or how to do it, where you could invent your own rules and build your own forms. Stated in terms which evoke the characteristic American restlessness, boldness, vision, and enterprise that pioneered the West—part adolescent arrogance and part courage and imagination—one discovers a fundamental American ethos behind this radical media movement.

Guerrilla Television Defined

The term "guerrilla television" came from the 1971 book of the same title by Michael Shamberg. The manifesto outlined a technological radicalism which claimed that commercial television, with its mass audiences, was a conditioning agent rather than a source of enlightenment. Video offered the means to "decentralize" television so that a Whitmanesque democracy of ideas, opinions, and cultural expressions—made both by and for the people—could then be "narrowcast" on cable television. Shamberg, a former *Time* correspondent, had discovered video was a medium more potent than print while reporting on the historic "TV as a Creative Medium" show at Howard Wise Gallery in 1969. Banding together with Frank Gillette, Paul Ryan, and Ira Schneider (three of the artists in the show), among others, they formed Raindance Corporation, video's self-proclaimed think-tank equivalent to the Rand Corporation. Raindance produced several volumes of a magazine called *Radical Software*, the video underground's bible, gossip sheet, and chief networking tool during the early Seventies. It was in the pages of *Radical Software* and *Guerrilla Television* that a radical media philosophy was articulated, but it was in the documentary tapes which were first shown closed-circuit, then cablecast, and finally broadcast that guerrilla television was practiced and revised.

Virtuous Limitations

Before the federal mandate required local origination programming on cable and opened the wires to public access, the only way to see guerrilla television was in "video theaters"—lofts or galleries or a monitor off the back end of a van where videotapes were shown closed-circuit to an "in" crowd of friends, community members, or video enthusiasts. In New York, People's Video Theater, Global Village, the Videofreex, and Raindance showed tapes at their lofts. People's Video Theater was probably the most politically and socially radical of the foursome, regularly screening "street tapes" which might include the philosophic musings of an aging, black, shoe shine man or a video intervention to avert street violence between angry blacks and whites in Harlem. These gritty, black-and-white tapes were generally edited in the camera, since editing was as yet a primitive matter of cut-and-paste or else a maddeningly imprecise back space method of cuing scenes for "crash" edits. The technological limitations of early video equipment were merely incorporated in the style, thus "real-time video"—whether criticized for being boring and inept or praised for its fidelity to the cinema vérité ethic—was in fact an aesthetic largely dictated by the equipment. Video pioneers of necessity were adept at making a virtue of their limitations. Real-time video became a conscious style praised for being honest in presenting an unreconstructed reality and opposed to conventional television "reality," with its quick, highly-edited scenes and narration—whether stand-up or voice over—by a typically white, male figure of authority. When electronic editing and color video became available later, the aesthetic adapted to the changing technology, but these fundamental stylistic expectations laid down in video's primitive past lingered on through the decade. What these early works may have lacked in terms of technical polish or visual sophistication they frequently made up for in sheer energy and raw immediacy of content matter.

With cable's rise in the early Seventies came a new stage in guerrilla television's growth. The prospect of using cable to reach larger audiences and create an alternative to network TV proved a catalytic agent. Video groups sprang up across the country, from rural Appalachia to wealthy Marin County, even to cities like New Orleans where it would be years before cable was ever laid. TVTV, guerrilla television's most mediagenic and controversial group, was formed during this time. Founded by *Guerrilla Television's* Michael Shamberg, TVTV produced its first tapes for cable, then went on to public television, and finally, network TV. TVTV's rise and fall traces a major arc in guerrilla television's history.

Shamberg had been thinking about getting together a group of video freaks to go to Miami to cover the 1972 Presidential nominating conventions. The name came to him one February morning while doing yoga at the McBurney Y in New York. He realized instantly that Top Value Television—"you know, like in Top Value stamps"—would also read as TVTV. He and Megan Williams joined with Allen Rucker and members of Ant Farm, the Videofreex, and Raindance to form TVTV's first production crew. Shamberg got a commitment from two cable stations and raised $15,000 to do two, hour-long tapes. The first, a video scrapbook of the Democratic Convention titled *The World's Largest TV Studio*, played on cable and would have been the last of TVTV were it not for an unprecendented review in the *New York Times* by TV critic John O'Connor who pronounced it "distinctive and valuable." With that validation, Shamberg was able to raise more money and hold the cable companies to their agreement, going on to cover the Repubican Convention the following month. *Four More Years* was the result; it is one of TVTV's best works, demonstrating the hallmarks of their iconoclastic, intimate New Journalism style.

Unlike the Democratic convention, chaotic and diffuse, the Republicans had a clear, if uninspired scenario to re-elect Richard Nixon. Instead of pointing their cameras at the podium, TVTV's crew of 19 threaded their way through delegate caucuses, Young Republican rallies, cocktail parties, anti-war demonstrations, and the frenzy of the convention floor. Capturing the hysteria of political zealots, they focused on the sharp differences between the Young Voters for Nixon and Vietnam Vets Against the War, all the while entertaining viewers with foibles of politicians, press, and camp followers alike. One Republican organizer's remark to her staff, "The balloons alone will give us the fun we need," epitomizes the zany, real-life comedy TVTV captured on tape.

Interviewed on the quality of convention coverage are press personalities whose off-the-cuff remarks ("I'm not a big fan of advocacy reporting."—Dan Rather; "What's news? Things that happen."—Herb Kaplow; "Introspection isn't good for a journalist."—Walter Cronkite) culminate with Roger Mudd playing mum's the word to Skip Blumberg's futile questions.

Punctuating the carnival atmosphere are venomous verbal attacks on the anti-war vets by onlookers and delegates who charge them with being hopheads, draft dodgers, and unpatriotic—a chilling reminder of the hostility and tragic confrontations of the Vietnam era.

TVTV follows the convention chaos, editing simultaneous events into a dramatic shape that climaxes when delegates and demonstrators alike are gassed by the police. Leavened with humor, irony, and iconoclasm, *Four More Years* is a unique document of the Nixon years. In it TVTV demonstrated journalistic freshness, a sardonic view of our political process and the media that covers it, and a sure feel for the cliches of an American ritual.

Forging A Distinctive Style

In forging their distinctive style, TVTV avoided voice-overs like the plague; they experimented with graphics, using campaign buttons to punctuate the tape and give it a certain thematic unity; and they deployed a wide-angle lens, which distorted faces as editorial commentary. The fisheye look, used at first out of practical necessity, since the Portapak lens often didn't let in enough light and went out of focus in many shooting situations, became a TVTV signature which led to later charges of exploitation of unsuspecting subjects. But in the beginning, it was all new and fresh and exciting. The critics pronounced that TVTV had covered the conventions better than all the networks combined, proving the alternate media could beat the networks at their own game and, as one pundit quipped, "for the money CBS spent on coffee."

Although the networks had ENG (electronic news gathering) units at the convention, the contrast was striking. Only a beefy cameraman could withstand the enormous apparatus, including scuba-style backpack to transport so-called "portable" television cameras. Fully equipped, they looked more like moon men than media makers. Compared to this, the light-weight, black-and-white Portapak and recorder in the hands of slim Nancy Cain of the Videofreex looked like a child's toy, which was part of the charm since no one took seriously these low-tech hippies. In video's early days, many didn't believe the tape was rolling because it didn't make the whirring sound of the TV film cameras, and much unguarded dialogue was captured because the medium was new and unfamiliar.

Television Enters the Picture

Thus established, TVTV went on to make their next "event" tape, but now for the TV Lab at PBS's WNET in New York. They were not the first to flirt with "Television." After the Woodstock Nation caught the networks' attention in 1969, the Videofreex were hired by CBS to produce a pilot which failed spectacularly in winning network approval. In 1970 the May Day Collective shot videotape at week-long anti-war demonstrations in Washington for NBC News although none of it was ever broadcast. The networks did air some news-breaking Portapak tapes, such as Bill Stephens's 1971 interview with Eldridge Cleaver over the split in the Black Panther party, shown on Walter Cronkite's *Evening News*. They were willing to overlook the primitive quality of tape which had to be shot off a monitor with a studio camera if it meant scooping their competitors, but the 1960 network ban on airing independently-produced news and public affairs productions remained in force, and any small-format tapes broadcast were usually excerpted and narrated by network commentators, beyond the editorial reach of their makers.

The introduction of the stand-alone time base corrector in 1973, a black box that stabilized helical scan tapes and made them broadcastable, changed everything. It was finally possible for small-format video to become a stable television production medium, which not only paved the way for guerrilla television to reach the masses but also for the rise of ENG and, eventually, all-video television production. Given TVTV's unprecedented success with *Four More Years*, it was only logical that they produce the first half-inch video documentary for airing on national public television.

The tape was *Lord of the Universe*, and its subject was the 15-year-old guru Maharaj Ji. Millenium '73, a gathering of the guru's faded flower children followers, was scheduled for the Houston Astrodome, which the guru promised would levitate at the close (like the Yippies at the Pentagon in '67, the guru knew how to create a media event). Elon Soltes, whose brother-in-law was a would-be believer, followed him with Portapak from Boston to Houston while other TVTV crew members gathered in Houston to tape the mahatmas and the premies (followers), getting embroiled in what was to be the most successful TVTV tape but also the most shattering for its makers. Fearful of mind-control and violence (a prankish reporter had been brained by a guru body guard not long before) and stricken by the sight of so many of their own generation lost and foundering in the arms of this spiritual Svengali, TVTV was determined to expose the sham and get out unscathed. The tape was the zenith of TVTV's guerrilla TV style.

Switching back and forth between the preparation for the actual onstage "performances" of the guru, cameras focused on "blissed out" devotees pathetically seeking stability and guidance in the guru's fold. Neon light, glitter, and rock music furnished by the guru's brother (a rotund rip-off of Elvis Presley) on a Las Vegas-styled stage was the unlikely backdrop for the guru's satsang or preaching to his followers. Outside, angry arguments between premies and Hare Krishna followers and one bible-spouting militant fundamentalist expose the undercurrent of violence, repression, and control in any extremist religion. TVTV cleverly played off two Sixties radicals against each other. Having traded in his role of counter-cultural political leader for that of spokesman for an improbable religion, Rennie Davis sings the guru's praises as Abbie Hoffman, one of guerrilla TV's Superstars, watches Davis on tape and comments on his former colleague's arrogance and skills as a propagandist. "It's different saying you've found God than saying you know his address and credit card number. . . " Hoffman quips, emphasizing the grasping side of this so-called religion.

Much in evidence is TVTV's creative use of graphics, live music, and wide-angle lens shots. As always there is humor leavening what was for TVTV a tragic situation. At one point our Boston guide to the "gurunoids" innocently remarks, "I don't know whether it's the air conditioning, but you can really *feel* something." The humor is a black humor, rife with an irony that dangerously borders on mockery but is checked by an underlying compassion for the desperation of lost souls. At home in the world of spectacle and carnival, ever agile in debunking power seekers, TVTV admirably succeeded in producing a document of the times that remains a classic.

Film's Hidden Impact

Paul Goldsmith, a well-known 16mm vérité cameraman, had joined TVTV along with Wendy Appel and was the principle cameraman on this and subsequent tapes, shooting one-inch color for the time in the Astrodome. Appel, also trained in film but an accomplished videomaker herself, would become TVTV's most versatile editor. Not surprisingly, some of the most critical people in creating the TVTV style came out of film: Stanton Kaye and Ira Schneider, who worked on the convention tapes, were both filmmakers as were Goldsmith and Appel. TVTV's raw vitality was a video and cultural by-product, but its keen visual sense and editing was borrowed, in large measure, from film.

TVTV won the DuPont-Columbia Journalism Award for *Lord of the Universe*, and, not long after, a lucrative contract with PBS to produce a series of documentaries for the TV Lab. *Gerald Ford's America, In Hiding: Abbie Hoffman, The Good Times Are Killing Us, Superbowl*, and *TVTV Looks at the Oscars* were made in the next two years. Some were equal to the TVTV name, like "Chic to Sheik," the second of the four-part *Gerald Ford's America*. But others showed a decline as the diverse group of video freaks who had once converged to make TVTV a reality—all donating time, equipment, and talent to make a program that would show the world what guerrilla television could do—began to stray in their own directions, no longer willing to be subsumed in an egalitarian mass, no longer able to support themselves on good cheer and beer. With the broadcast of *Lord of the Universe* some of the best minds in guerrilla television unwittingly abandoned their utopian dream of creating an alternative to network television. Their hasty marriage with cable was on the rocks when TV—albiet public television—seduced them with the fickle affection of its mass audience.

The Beginning of the End

In 1975, TVTV left San Francisco, which had been home-base during its halcyon days, for Los Angeles. This move proved pivotal. They had a contract to develop a fiction idea for the PBS series "Visions." This was not so much a departure from TVTV's orientation as it might seem. They had been mixing fictional elements in their documentary tapes all along, the most notable being the Lily Tomlin character in the *Oscars* show. TVTV's style had been modeled on New Journalism and the flamboyant approaches of writers like Hunter Thompson, of Gonzo Journalism fame, who wrote nonfiction like it was fiction.

Supervision consisted of a number of short tapes, "filler" to round-off the "Visions" series' hour. It traced the history of television from its early days in the labs of Philo T. Farnsworth to the year 2000 and an imagined guerrilla take-over of a station not unlike CNN. Forsaking the video documentary form that they had poineered caused some internal battles, but it wasn't until their pilot for NBC, *The TVTV Show*, that the end was in sight.

Part of the problem was that TVTV knew how to make a video documentary—in a way, they had invented it—but they didn't know the first thing about producing comedy for "Television." In documentary shooting, improvisation on location was TVTV's trademark; the primitive and evolving nature of portable video equipment and the unpredictable power centers which were TVTV's main targets demanded an adaptive and creative attitude toward all new situations, something TVTV excelled at. But shooting actors in a studio with a set script that never equalled the humor of their documentary "real people" demanded a whole new expertise which TVTV realized too late they couldn't afford to invent as they went along.

Another part of the problem was that so long as TVTV was making documentaries, the group had its original focus. Once they began making entertainment for mass audiences, their once-radical identity and purpose was gone. For some, the evolution was a gradual and acceptable one. After charges of "checkbook journalism" over the ill-fated interview of Abbie Hoffman, who was then a fugitive, Shamberg lost some of his journalistic zeal. Harsh criticism of the treatment of Cajuns in *The Good Times Are Killing Me* further tarnished TVTV's reputation. With people like Bill Murray and Harold Ramis (who would later become celebrities on "Saturday Night Live") eager to work with TVTV, the lure of collaborating with talented actors in an area removed from journalistic criticism, funding battles, and the pressures of producing documentaries for public TV was certainly appealing. But for others who still believed in the dream of changing television, the decision proved a hard one because it meant the dream was dead. And with it went the all-for-one spirit that knitted together their disparate egos: TVTV no longer had the fire and purpose it needed to weather the rough storm of a mid-seventies transition.

It took a few years as TVTV paid off its debts before their official demise. In the meantime, Shamberg, who had seen the end coming, was already preparing his next venture. He bought the rights to the Neal and Carolyn Cassady story and produced the film *Heartbeat*. It was a box office flop, but convinced him to go on. In 1983, two films later, he produced the Academy Award nominee *The Big Chill*, a reunion film about a group of late Sixties' hippies who meet at the funeral of one of their own and reflect on how they've changed and been affected by "the big chill." Although the film was based on director/writer Larry Kasdan's friends, it could have been about TVTV.

Changing Times

The fact that TVTV changed along with their times should come as no surprise. TVTV wasn't the only group to pull apart during the late Seventies. The media revolutionaries were growing older and changing—assuming responsibilities for marriages, homes, and families—living in a different world from the one which once celebrated the brash goals and idealistic dreams of guerrilla television. The promise that cable TV would serve as a democratic alternative to corporately-owned television was betrayed by federal deregulation and footloose franchise agreements. Public television's early support for experimental documentary and artistic work in video slowed to a virtual halt—the sad demise of WNET's TV Lab in recent months its latest instance. And funding sources that had once lavished support and enthusiasm on guerrilla TV groups now turned a cold shoulder, preferring to support individuals rather than groups and work that stressed art and experimentation rather than controversy and community.

Once the possibility of reaching a mass audience opened up, the very nature of guerrilla television changed. No longer out to create an alternative to television, guerrilla TV was competing on the same airwaves for viewers and sponsors. As the technical evolution speeded up, video freaks needed access to more expensive production and post-production equipment if they were to make state-of-the-art tapes which were "broadcastable." Although some continued making television their own way, pioneering what has since become the world of low power TV and the terrain of public access cable, many others yearned to see their work reach a wide audience. Without anyone's noticing it, the rough vitality of guerrilla TV's early days was shed for a slicker, TV look. The "voice of God" narrator, which had been anathema to TVTV and other video pioneers, was heard again. Gone were the innovations—the graphics, the funky style and subjects, the jousting at power centers and scrutiny of the media. Gone was the intimate, amiable cameraperson/interviewer which was a hallmark of alternative video style. Increasingly, video documentaries began looking more and more like "television" documentaries, with stand-up reporters and slide-lecture approaches that skimmed over an issue and took no stance.

Where one could see the impact of guerrilla television was in its parody. Sincere documentaries about ordinary people had been absorbed and transformed into mock-u-entertainments like "Real People" and "That's Incredible!" The video vérité of the 1976 award-winning *The Police Tapes*, by Alan and Susan Raymond, had become the template for the popular TV series "Hill Street Blues." In the sixties, Raindance's Paul Ryan proclaimed, "VT is not TV," but by the Eighties, VT *was* TV.

Today, in an era of creeping conservatism, the ideals of guerrilla television are more in need of champions than in its heyday when it was easier to stand up for a democratic media that would tell it like it is for ordinary people living in late 20th century America. Few have come along to take up the challenge of guerrilla television's more radical and innovative past. Although the collectives with names like rock groups—Amazing Grace, April Video, and the Underground Vegetables—have long since disappeared, many notable pioneers continue to keep alive their ideals, some working in public access cable like DeeDee Halleck (of Paper Tiger Television), or from within the networks like Ann Volkes (an editor at CBS News) and Greg Pratt (a documentary video producer for a network affiliate in Minneapolis), or as independent journalists like Jon Alpert (freelance correspondent for NBC's "Today Show") and Skip Blumberg (whose portraits of Double Dutch jumpers and Eskimo atheletes still appear on public television). But a younger generation eager to draw from this past to forge a new documentary video future have yet to appear on the horizon. Either they are discouraged by the lack of funding and distribution outlets for innovative and/or controversial work and a cultural milieu content with the new conservatism, or they are unaware of the past and unconcerned about the future. The goal is not to recreate that past—no one really wants to see the shaky, black-and-white, out-of-focus, wild shots that suited the primitive equipment and frenzy of video's Wonder Bread years; the goal is to recapture the creativity, exploration, and daring of those formative years. Perhaps the technology and the burning need to communicate and invent forms will prevail. Independents with Beta and VHS equipment have been documenting the struggles in Central America. Lost amid the home video boom, a new generation of video guerrillas may be in training yet.

McLuhan's view that "the medium is the message" was embraced then rejected by the first video guerrillas who asserted that content *did* matter; finding a new form and a better means of distributing diverse opinions was the dilemma. That dilemma is still with us. How a new wave of video guerrillas will resolve it and carry on that legacy, human and imperfect as it may be, should prove to be interesting and unexpected. More than guerrilla television's future may depend on it.

FAMOUS YINYAN SCHOOL
BANG
Please keep
The New York Times

MEET THE PRESS: ON *PAPER TIGER TELEVISION*

Martha Gever

A COMMUNICATIONS REVOLUTION is underway, or so we are constantly told by our media. The movement behind that revolution is the unprecedented growth in electronic communications technology, developments which indeed are changing the world but not in ways usually associated with revolutions. When most of us switch channels on the TV set we can predict with great accuracy what we will encounter there. Depending on the hour and day of the week we anticipate soap operas, news, comedies, sports, cop shows, cultural programs, talk shows, and the like. Of course, the offerings for subscribers to cable television services are more diverse, or are they?

Diversity was the revolutionary promise of cable TV—the potential of more than 100 channels serving as many specialized audiences. Structurally, cable TV is frequently compared to print media: newspapers, magazines, and other journals catering to various interests and needs. In the same breath, cable TV is contrasted with broadcast television and radio where the differences between stations are limited by a finite spectrum of electromagnetic frequencies; in other words, relatively few channels can be used within a broadcast area. Yes, there are more programs on cable TV than Marconi ever dreamed of, but hooking into the system usually brings few surprises and little innovation.

In fact, the comparison of cable television and the press is only valid in a technical sense. The required capital investment and high operating costs of cable systems dictate monopoly control within a given geographic area; in each locale a franchise is awarded to a single cable operator. The resulting structure closely resembles telephone networks or gas and electric utilities, and when profit-making becomes a factor conflicts between public and private interests arise. The mechanisms developed to mediate such conflicts are—as in the case of public utilities—government regulatory agencies at municipal, state, and federal levels. The most powerful of these, the Federal Communications Commission, however, has been moving steadily towards deregulation for years, and under the

Reagan administration the identification of public interests with business interests seems complete.

Deregulation advocates base their arguments on a private enterprise model of communications: programs on different TV channels constitute a range of commodities, and the public is defined as a group of individual consumers.[1] Likewise, lobbyists for the cable industry assume the same position, claiming their range of commodities is almost unlimited and, therefore, like publishers, cable TV should be immune from government interference.[2] In a recent issue of *Screen* Nicholas Garnham disputes the prevailing capitalist interpretation of public interests in relation to television and the government's role in protecting those interests:

. . . the truth is that while the public regulation of broadcasting has been legitimized *in terms of frequency scarcity, its* justification *lies in its superiority to the market as a means of providing all citizens, whatever their wealth or geographic location, equal access to a wide range of high quality entertainment, information and education, and as a means of ensuring that the aim of the programme producer is the satisfaction of a range of audience tastes rather than only those tastes that show the largest profit.*[3]

Garnham proceeds to an analysis of the political meaning of the marketplace definition:

. . . we are here in the presence of ideology in its pure, classical form; that is to say a social analysis that not only misrepresents its object of analysis by focusing on its surface rather than its underlying structure and by denying its real history, but also misrepresents it in such a way as to favor the interests of the dominant class. In this case the trick is played by concentrating upon the technical potentialities rather than upon the social relations that will determine the form in which those potentialities are realized and by denying history by exaggerating the novelty of the process in question.[4]

Certainly, the "process in question," i.e. TV, affects the nature of communications, the exchange of information, but the end result is not necessarily increased freedom or even increased information available to the public. So far, in this country the benefits of advanced technology have accrued mainly to investors in information industries, the corporate clients of these investors in information industries, the corporate clients of these industries, and, not incidentally, to the military.[5] In an information economy this trend leads to an increasing concentration of wealth *and* information.

Despite the flaws in the cable operator's equation of cable TV and print media, the analogy is applicable when considering the problem of concentrated ownership. The number of daily newspapers published in the U.S. has declined drastically; between 1923 and 1976 the total decreased by 271,[6] and presently most medium and small U.S. cities have only one daily paper. Granted, this drop can be partially attributed to the popularity of television as a source of news and other types of information. This fact does not explain, however, the continued expansion of large newspaper chains like Knight Ridder, Gannett, and Hearst, which have increased their spheres by buying up their poorer competitors.[7] Predictably, the style and content of journalism and the political outlook found in these papers has become homogenous. The same kind of empire building is now occurring in cable TV, once the domain of local entrepreneurs but now largely controlled by giants: Telecommunications, Inc. (3,300,000 subscribers); American Television Communications, owned by Time, Inc. (2,400,000); Group W Cable, Westinghouse Broadcasting and Communications, Inc. (2,007,000); and the list goes on.[8] Other entertainment industrialists like Gulf and Western, the parent company of Universal Pictures; the broadcast networks; and Playboy are now active in the cable programming scene. No wonder cable fare is indistinguishable from what's on CBS or PBS.

*

Returning, then, to the prospect of flipping the dial, TV viewers hardly expect to find a program which criticizes media monopolies and analyzes the politics of the communications industry. But if those viewers live in Manhatten they can watch such a program—*Paper Tiger Television*, a weekly series on public access cable channels C and D.

The mandatory provision of public access to cable programming is one of the few vestiges of a non-commercial, public service concept of communications—borrowing a phrase from Fred Glass, a crack in the TV tube.[9] Access producers around the country often define their work as grassroots, democratic television (as opposed to the notion of consumer, marketplace democracy or the false push-button democracy of QUBE, for example). Initially produced as part of the *Communications Update* series,[10] *Paper Tiger Television* attacks the ideology and economics of mass media in the context of the industry itself. "The power of mass culture," the *Paper Tiger* producers state, "rests on the trust of the public. This legitimacy is a paper tiger." Thus the series' title.

I use the plural when referring to the *Paper Tiger* producers because the three-year old series is collectively made. Membership is fluid, but the core group includes DeeDee Halleck, who conceived the series and continues to play a major role, Diana Agosta, Pennee Bender, Skip Blumberg, Shulea Cheang, Esti Marpet, Caryn Rogoff, David Shulman, Martha Wallner and Alan Steinheimer. Their program design is simple: each week a different publication(s) is read, i.e., analyzed, by a different commentator. Although most of these critics could be classed as intellectuals, it's a motley group: communications theorists, visual artists, writers, economists, musicians, psychiatrists, lawyers, etc. Unlike a conventional talk show, what is read and what is said is left entirely to the featured commentator. The series is not intended as a display of professional, specialized expertise, but rather as an invitation to develop critical consciousness.

While *Paper Tiger* programs have been shown in college classrooms and numerous other closed-circuit screenings, the shows originate as television programs. Therefore, the *Paper Tiger* producers try to cultivate an audience; their strategy mixes consistency—a regular time slot and basic format—with provocative, eye-catching productions—no panel discussions in front of a blue curtain. Although programs are taped and repeated periodically, most shows premier as live performances, creating a refreshing spontaneity as well as a lot of rough edges.

What would be considered unprofessional on most TV programs—for instance, the "talent" talking to the camera operator, a misdirected camera pan—is acceptable on *Paper Tiger TV*. Slick, seamless productions are consciously avoided in favor of obviously low-budget, even amateurish simplicity. A paper movie, a continuous strip of paper moved by hand-cranked rollers, displays titles and credits. Or hand-lettered cards are held in front of the camera. Sets are often flat, cartoon-like cloth backdrops—the interior of a graffiti-painted New York subway car, a homely room with flowered wallpaper, a gray basement wall adorned with pipes and meters.

The resulting appearance of *Paper Tiger* shows reflects both practical and conceptual decisions. Access producers in New York City must pay for studio time at the rate of about $55.00 per hour (in some communities access to equipment, studio space, and editing facilities is free). The *Paper Tiger* crew rents their studio for one hour per show, which allows only 30 minutes for preparation. Even with this economy, rental fees comprise the major budget item; the average cost of a *Paper Tiger* program is $200, a negligible amount compared with $150,000 spent on a typical half-hour commercial television program. Simply stated, bargain-basement production values make the series possible. Aside from these material concerns, the overt style of *Paper Tiger's* design reinforces its purpose—to reveal, not conceal, the working of media production. The flat sets, punkish graphics, and the cost-breakdown given at the close of each show implicitly expose invisible, finely-tuned methods of media seduction. This is Brecht's "alienation effect" applied to television:

The A-effect consists of turning an object of which one is to be made aware, to which one's attention is to be drawn, from something ordinary, familiar, immediately accessible, into something peculiar, striking and unexpected.[11]

In *Paper Tiger Television*, the object drawn to one's attention is mass media.

*

There are, to date 70 *Paper Tiger* programs. While the general framework I have described gives the series a unity, each half-hour show is tailored to its particular subject and views of the particular commentator. And since *Paper Tiger Television* is not an abstract idea but an actual set of programs, I will examine some of the parts that make up the whole. In this review of *Paper Tiger TV* I will necessarily neglect or only mention details of particular programs, but that should not be mistaken for negative judgment. Instead my purpose is to evaluate the series as an integrated project, and I have chosen my examples accordingly.

As I noted earlier, the majority of the invited critics are intellectuals, and many of them are university teachers. The programs, therefore, often create the atmosphere of a lecture hall, but a very unusual university-of-the-air. Herbert Schiller, whose six consecutive readings of the *New York Times* initiated the series, offers his audience a lesson in communications theory without pedantry. His ability to present a barrage of information in an entertaining manner depends both on the clarity of his reasoning and on his specific revelations about the powerful politics of the world's most influential newspaper. Just as in his books, which include *The Mind Managers* and *Who Knows? Information in the Age of the Fortune 500*,[12] Schiller doesn't mince words in his analysis of the economic interests which shape the form and content of this important source of information.

In the program devoted to the Sunday edition of the *New York Times*, Schiller holds up the four pound document, all 712 pages, and points out that three-quarters of the space in the first section is occupied by ads. Turning to the magazine section, he describes the contents of the features as well as the omissions: the story on New York City architecture doesn't discuss the role of real estate interests; the story on the high cost of medical education in the U.S. ignores the economics of the medical system; a third story, on mining rubies in Thailand fails to mention U.S. involvement in Thai politics. What the *New York Times* offers its readers, Schiller concludes, is fragmentary, depoliticized accounts of events and issues, limited knowledge which hinders thoughtful, informed evaluation. The high advertising content indicates the tacit capitalist slant of the *Times*.

This attention to advertising as an integral part of a publication's content is common in *Paper Tiger TV* programs. Tuli Kupferberg, reading *Rolling Stone*, notes a preponderance of what he calls "get drunk ads."

Myrna Bain refers to the numerous ads for liquor and cigarettes in *Ebony* as "ads which are meant to relieve our pain." Ads in both the *Wall Street Journal* and *Working Woman* read like the articles on facing pages: images of successful businessmen in the *Wall Street Journal* or successful businesswomen pleasing male bosses in *Working Woman*. And as Martha Rosler leafs through *Vogue*, the ads and the feature articles become indistinguishable. Schiller encounters the same confusion in the fashion pages of the *Times* Sunday magazine, and even in the book review section the relationship between the major publishers' ads and the books reviewed is evident.

In another half-hour segment Schiller discusses the reporting of international events in the *New York Times*, which he summarizes as "incomplete and deceptive." He faults the *Times* for selectively ignoring contextual information in the guise of factual reportage. Schiller's central example of inaccurate news coverage concerns the very mechanisms which produce the news. At issue is the UNESCO International Commission for the Study of Communications Problems; the commission's report, commonly known as the *McBride Report*, named after the commission's chair, Sean McBride;[13] and the subsequent hostility of the U.S. press in general and the *New York Times* in particular to that report.[14] In brief, the *McBride Report* advocates a need for a New World Information Order, a restructuring of international communications now dominated by U.S. and European media to the detriment of developing countries in Asia, Latin America, and Africa. The almost unanimous negative response from mainstream U.S. media has created a specter of freedom of the press endangered by political propaganda.[15] But, Schiller asks, can the *New York Times* report objectively on proposals to change the international flow of information when they are implicated in that debate? In his analysis, the answer is no.

Against the background of Schiller's exposition of the controversies surrounding proposals for a New World Information Order, subsequent *Paper Tiger* programs examine the present world information order. *Business Week*, as Harry Magdoff points out, is read almost exclusively by business executives. Nor surprisingly, then, a common topic covered in the magazine is the pursuit of profits. As an example of the social consciousness purveyed in the pages of *Business Week*, Magdoff reads excerpts from an editorial on unemployment—not industrial unemployment, but that of middle managers who have been phased out of their jobs due to advanced technology.

Like Schiller, Magdoff's delivery is direct and articulate. He connects *Business Week's* place in the McGraw-Hill publishing empire—a vast one which is also the leading data-base publishing company in the U.S.—to its function as a definitive voice of corporate ideology. Magdoff, the editor of the socialist *Monthly Review*, even recommends reading *Business Week* as a guidebook to corporate strategies.

Despite Magdoff's lucidity, a non-stop monologue cannot usually sustain a half-hour program intended for a popular audience. The *Paper Tiger* producers are aware of this, and they interrupt and enhance his critical reading with a raft of statistics about the magazine, its subscribers, and its parent company, displayed against shots of workaday Manhattanites and skyscrapers. All this is set to music: the new wave version of the Beatles's song *Money*. This kind of collage data, visual commentary, and ironic music is common to many *Paper Tiger* productions—a sort of anti-commercial break.

A further level of interpretation is also edited into Magdoff's critique—the opinions of some *Business Week* readers presented in man-on-the-street interviews (they are all men, as are most *Business Week* subscribers). These remarks seem rather shallow, however, reiterating rather than augmenting Magdoff's observations. This device proves more successful in the *Ebony* program, where an interviewee discloses why she doesn't read that magazine—except in the doctor's waiting room. And contrasted to the stereotypes of black children which critic Myrna Bain finds in *Ebony*, interviews with a group of children refute those images.

Contradiction and its close relative, ironic humor, are the twin scalpels consistently wielded by *Paper Tiger* critics and producers. Sometimes powerful contradictions are rendered verbally in redefinitions of a publication's substance and impact. That's Schiller's tactic. So, too, the political scientist Archie Singham provides an unorthodox explanation for the prestige and authority of specialized journals on foreign policy:

Intellectuals are involved in the task of obscuring reality because we earn our living by being obscure. . . . Obscurity is the essence of the scientific effort to destroy the capacity of people to understand reality in a fundamental sense. . . . Ordinary citizens feel that Foreign Affairs *is not their business.*

Turning the glossy pages of *Vogue* magazine, Martha Rosler starts with an exaggerated parody of television commercials for elegant products:

What is Vogue? *What is fashion? It is glamor, excitement, romance, drama, wishing, dreaming, winning, success. It is luxury, allure, mystery,* romance, *excitement,* Love, *splendor. . .*

Gradually, her mock-promo shifts to direct criticism:

What is Vogue? *It is photography, it is voyeurism, it is mystification, it is fascination, desire, and identification. . . . It is self as object, as sculpture, as creation. It is submissiveness in the guise of witch power over men, over women, over careers, over the private world.*

Rosler not only speaks contradictory language, she also gives these contradictions a face. As a final act, she carefully applies make-up—

mascara, eyeliner, rouge, lipstick, etc.—enacting the kind of personal transformation proposed on the pages of *Vogue*. When finished with her work, she holds up a mirror which reflects a divided visage—a sort of before-and-after comparison combined in one image. Half her face is artificial, the other half pale, but less garish, and as a whole her mirrored complexion presents a strange, disturbing sight.

Rosler is a visual artist as well as a writer and teacher. In most other cases, visual contradictions are built into *Paper Tiger* programs by the producers. One of the simplest but, I think, most inspired examples of this is the oversize American flag draped behind Conrad Lynn as he describes the evolution of *Commentary* magazine—originally a liberal journal which is now a prominent organ of right-wing political opinion—a shift which Lynn attributes to *Commentary's* embracing the Republican Party in the 1960's.

In several *Paper Tiger* episodes the ironic *mise-en-scene* is elaborated by actors. Unlike the injection of unstudied naturalism achieved through street interviews, these characters are more like caricatures drawn from the pages of the publication under consideration. What results is often a living counterpoint to the featured critic's remarks. In this era when intricate video editing has become synonymous with the construction of comparative juxtapositions, this model of montage *within the video frame* seems particularly anachronistic and, I believe, more than just a stylistic choice. One primary difference between an edited sequence constructed according to Eisenstein's principles of dialectical montage and these *Paper Tiger* scenarios is the locus of production. A dialectical opposition and comparison of images and associations manufactured in the editing room is contingent on the conceptual skill of the editor/artist *and* the editing apparatus. The dialectical montage in *Paper Tiger* programs is basically theatrical, realized in production planning and through interactive performances. Once recorded, the performance is unchangeable (although a shot or scene might be re-recorded, and, at times, programs are amended in the editing room.) The significance of the implied rejection of technology is a demystification of television designed *for* television. Thus, this serial critique of print media is effected in a form critical of the greater mass media, TV.

Let me illustrate this point with a synopsis of the staging for one *Paper Tiger* show, the *Psychology Today* program. At the opening Carole King sings her familiar lyrics,". . .all you have to do is call, and I'll come running to see you again. . . ." A man is seated in front of a fixed camera, and a woman is lying on a couch positioned next to, and, from the camera's point of view, behind the man. The scenario immediately becomes recognizable: the psychoanalyst's office—in fact, a low-budget replica of Freud's office. The patient begins to pour forth her confessions;

meanwhile the shrink gets distracted, drums his fingers, checks his watch. Then, unexpectedly, he turns to face the camera, and, in the style of a theatrical aside, he commences his discussion of *Psychology Today*. This shrink is, in fact, a real but irreverant psychoanalyst, Joel Kovel, who proceeds to dissect his chosen text. While Kovel talks to the audience, the patient, played by Melissa Leo, chatters and sobs in the background. At last, the doctor addresses the patient: "Time's up." She writes a check for $75.00 and sincerely thinks him for his time.

Kovel's debunking of the mental health industry, represented on the pages of *Psychology Today*, set against a parody of psychoanalysis is ingenious, and perhaps the most successful instance of the montage technique I have described. But this scene also introduces problems concerning the representation of cultural stereotypes in human form. Leo's female persona is so emotionally obsessed and narcissistic that she never notices her shrink's total inattention. Another familiar and derogatory feminine type appears in the *Seventeen* show, painting her toenails while Ynestra King discourses on the socializing hype pandered by that magazine. While Seraphina Bathrick discusses the feminist and anti-feminist content of *Working Woman*, Leo mimes a vacuous secretary straight out of a network sit-com.

The social *tableaux vivants* devised by the *Paper Tiger* collective bring to mind the political theater of Brecht and contemporary groups like El Teatro Campesino and the San Francisco Mime Troupe, but so far the performances seen on *Paper Tiger TV* are more like shadow plays projected by mass media than dynamic images of social relations. Again, Brecht's "alienation effect" can be detected, and a passage from his description of acting techniques intended to "make the spectator adopt an attitude of inquiry and criticism," which outlines a more complex approach to characterization, might be instructive.

Because (the actor) doesn't identify himself[sic] with the character he can pick a definite attitude to adopt towards the character whom he portrays, can show what he thinks of him and invite the spectator, who is likewise not asked to identify himself[sic], to criticize the character portrayed.

The attitude which he adopts is a socially critical one. . . his performance becomes a discussion (about social conditions) with the audience he is addressing. He prompts the spectator to justify or abolish these conditions according to what class he belongs to.[16]

Another, more videoesque experiment, which avoids acting problems but uses other methods of matching subject matter with setting, was tried in Tom Weinberg's reading of the *Wall Street Journal*. The program was shot on location, following the typical route of a Wall St. executive on a typical weekday morning. Visually, the results are often delightful: Weinberg in respectable suit and fedora perusing his businessman's bible aboard a Westchester-to-Grand Central commuter train, close-up shots of his Wall St. uniform, Weinberg scanning the latest stock quotations jammed into a rush-hour subway car, concluding with Weinberg in front of the Stock Exchange itself. The camera work and editing of the piece are remarkable well-crafted, but the show falls flat. Unfortunately, Weinberg has very little to say beyond mere description, proving again that the strength of *Paper Tiger Television* ultimately depends on the content of the commentators' critiques.

The last *Paper Tiger* program I will examine here features Joan Braderman, who has a great deal to say about her media item, the *National Enquirer*. Braderman's clever intelligence and entertaining wit are essential ingredients in the program's composition, but two other aspects of her presentation make it a significant contribution to the *Paper Tiger* repertoire. First, like Martha Rosler reading *Vogue*, Braderman explores the irrational appeal of mass publications. In her words: "Everyone knows the *National Enquirer* is a rag, but it engages people's desire." As Braderman notes, the contradiction for *Enquirer* readers lies between the need for community, satisfied by the complicity in gossip offered them, and the fact that that community is mythical. Her discourse on the social pathology which produces *Enquirer* readers is as informative and insightful as her recitation of data concerning the *Enquirer's* production and distribution—to approximately 19 million people every week.

The second notable factor is formal—not form divorced from content but as the carrier of meaning. At first, the *National Enquirer* show seems to negate my previous characterization of *Paper Tiger TV* as the antithesis of high-tech media. In this program a variety of video techniques, including special effects, are combined to create a dynamic, disjointed collage.[17] Yet even in this foray into the wonderland of electronic manipulation, the medium is governed by the message. Take the use of special effects, the prime marker of technical wizardry: unlike the sophisticated, often spectacular image-processing seen in much contemporary video work, the effects here look crude. The awkward wipes and insets of Braderman's face with silly undulating borders have no pretense to good taste. Actually, the frames echo the flashy graphic style of the *Enquirer*. Braderman's analysis of the schizophrenia of U.S. culture is propelled by her confession of personal addiction mixed with a good dose of sarcasm; likewise, the animated scandal sheet look of the *Enquirer* show satirizes seductive state-of-the-art TV.

*

Paper Tiger Television could only be made in New York City, the publishing capital of the U.S. Only in New York could producers recruit the variously talented media critics their series requires. I'm not saying that individual shows can't be made elsewhere. They can, and one future projection for the series entails a distribution and production network. Nor do I mean that public access producers in other communities cannot borrow ideas or replicate production methods. Still, *Paper Tiger TV* is really a local show, grounded in a particular place. At the same time, *Paper Tiger TV* encourages its audience to consider global issues, to think about how communications industries affect their lives and their understanding of the world. Consciously mixing almost primitive video techniques with sophisticated ideas, adding humorous touches to enliven serious questions, *Paper Tiger TV* can be described as a 1980's version of Brecht's didactic theater. *Paper Tiger's* didactic television, like its antecedent, weds analytic processes to popular forms in order to reveal social relations and social inequities. And the purpose is not mere criticism. "This criticism of the world," as Brecht said, "is active, practical, positive."[18]

By using public access channels to challenge media systems controlled by corporate conglomerates, *Paper Tiger TV* proposes a real communications revolution.

Excerpts from program transcripts.

HERB SCHILLER READS THE *NEW YORK TIMES*

The *New York Times* occupies a very special place in American life and also in the life of the metropolitan New York area. . . The *New York Times* circulates in the center of the whole system of finance and international money-making and property owning. The *Times* is what the decision makers read. The *Times* is the paper of record. That means it's supposed to—it doesn't always do this, in fact, it rarely does it—but it's supposed to be able to have verbatim texts, to say nothing of legitimate public statements. Another crucial ingredient of the *New York Times* is that the stories that appear on its front pages invariably also appear as the leading stories in the newsrooms of the three major networks—ABC, CBS and NBC. So the *New York Times* does in fact occupy a very special place.

. . . And we may say that the *New York Times*, if we may give it one central function, is in this sense always—and I use the word always, which is a very, very tricky word to use because few things are *always*—

but the *New York Times* we can say always is concerned with and always will feature the long term interests, as *it* determines them, of this entire system. There will be occasions, of course, when this will come into collision with one or another sector of powerful interests in this society, and on those occasions some times there is confusion, but nevertheless, remember, it is the steering mechanism of the system overall and not for any one particular side of it.

Let us examine two of the central questions that confront us as individuals, our nation, and the world, two central questions upon which we can make any determination of how a source of information is behaving. . . Let us take as our text today's *New York Times*, October 28th, and let us see a crucial question discussed in the columns of the *Times*. Let us point, for example, to the story of the AWACS, about which we have been hearing endless things and reading endless amounts for days and days, and, as we know today, the AWAC sale has been approved. Let us see how the *New York Times* talks about that and how it presents it in a picture. We have on the front page of today's *New York Times* a very interesting double photograph. We have three or four senior senators looking at a score sheet, and then beneath it we have a detailed picture of that score sheet. What does that tell us? We are told in the adjoining column who was switching, what is going to be the vote, how many votes there are. The entire story is one almost as if we are talking about a golf card score. It's almost as if we're discussing the World Series. This is a game. Will he switch, or won't he switch?

On an inner column we have one of the *Time's* serious analysts, Hedrick Smith, saying "Will the President Maintain Momentum?" Again, a sporting metaphor. The entire issue is presented in terms of votes—vote switching, what will a representative or senator do? And finally, there is a mention in still a third article in the same issue of how one particular senator switched his vote because highly classified information was made available to that senator. We are not told what that highly classified information is.

So we come out of this tremendous story on the penultimate day of the passage of this legislation with three major stories, endless verbiage, two photographs, and what are we given? Just a tremendous amount of superficial, frothy detail of how people's votes are moving, why they're moving. Aren't they moving? Is there momentum? How does the scorecard rest? What are we told about the Middle East? What are we told about the fact that the United States is the main supplier of arms to the Middle East? What are we told about the role of Saudi Arabia as a protector and as the surrogate for the U.S.—the role that Iran played?

. . . Now let's turn to one last quick article in terms of standard of living questions and let's see how the *Times*—and here we have a long, rich historical record which doesn't have to come only out of October 28th—how does it treat our labor population? How does it treat our working population? On October 27th, the day before today, there was a very small article, on page 18 mind you, describing the possibility of a general strike in Philadelphia to defend the rights of the teachers. In even a lesser article in a much, much more obscure part of today's paper, we find on page 16 that we get a tiny box saying that a general strike has been averted, with no real indication that this has been a tremendous victory for organized labor in Philadelphia, coming at a time when organized labor is under tremendous pressure from a very powerful anti-labor administration.

MYRNA BAIN READS *EBONY*

MYRNA BAIN: We're looking at an August 1980 special issue of *Ebony* magazine. As you can see it says, "Blacks and the Money Crunch: What It Means to You and How to Survive in These Hard Times." Times are tighter than tight. "White Recession, Black Depression." It's tight times.

Ebony magazine was founded in November 1945 by a man named John Johnson, who was born in Arkansas in 1918. . . Mr. Johnson started out with a smaller size publication than this, *Negro Digest*, that he founded in 1942 in Chicago, but this is his heart and soul, and he has made it the heart and soul of 1,600,000 black Americans who buy this publication either by subscription or on the newsstands. In fact, 83% of *Ebony* is now sold through subscription.

The push and the emphasis of *Ebony* magazine is how to put your money where your soul is.

. . . This special issue touting *Ebony—Ebony*, the colored answer to *Life* magazine. This special issue will tell us how to make a million, as in Nathaniel West, a cool million. . .

WOMAN(sitting on her front steps): I used to read it a lot more when I had more illusions about life in general, but I only read it now if I go to the doctor's office or somebody else has it.

I think that black people struggled a lot to promote positive images of themselves, so we projected black pride, and I think that magazines like *Ebony* reflected that *then*. But I think that it's a mistake to keep projecting "We've made it; we've made it," when it's only a minute number of black people who've made it. For instance, computer programming—no schools as far as I know in neighborhoods like East New York or Crown Heights even have computers. Our children are getting an obsolete education. By the time they get to the job market, they won't have a

marketable skill. They won't be prepared for the real world. How can you compete with someone who's being prepared to deal with the real world? How can you say that you have the same opportunity? You don't. And a lot of people read this magazine and people go around feeling that something's wrong with *them* because they didn't make it—because they can't accomplish all the things that these people in the success stories have.
MYRNA BAIN: We have another special issue. This is August 1974, the issue on the black child. Now August is a very important time to *Ebony*. It's their special issue month. Every August, since about 1951, they have run a special issue on money, on children, on the black middle class—a special issue. August is Emancipation Month, not only for blacks in this country but throughout the English-speaking world in this hemisphere. Slavery would be abolished in August and then paper independence would be granted in August. It's also the time when the sugar cane crop didn't need to be brought in anymore; it was in already. So we got freed in August when nobody didn't need us to work.

. . . This is the 1982 November issue featuring Gladys Knight and her brother Bubba. "Black Dynasty: Texas Family Builds Oil and Cattle Empire," "Should a Woman Make the First Move?" "Should you Marry Someone with Children?" "Eldricks Tiger Woods: A Golfing Sensation at Six." You, too, would pay $1.75 for this with Ronald Reagan in office in November 1982. You would buy this. Obviously, there are burning issues in here.

JOAN BRADERMAN READS THE *NATIONAL ENQUIRER*

Welcome to the world of the National Enguirer. *This is it. The contemporary '80s locus of the Great American contradiction. It's a fucking veil of tears for most of us folks, but the* Enquirer *like the* Readers' Digest *is upbeat. Ronnie and the rich are getting richer. They're buying super nukes, super dicks, those deaths. And the poor? Well, there are a hell of a lot more of us and we're trying to organize in the belly of the monster. But that bad news is that we're freakin' out. In the third world they just ain't going to take it no more, but the American press says that here at home life is just swell. Hey, we're reborn, Wallace conservatives, E.T. lovin' voyeurs, and you guessed it—our paper of free democratic choice—*The National Enquirer.
The *Enquirer* used to be a temperance magazine in the '20s. When Gene Pope bought it in '52, he borrowed $75,000 and turned it into this mass circulation weekly. It was his idea to get it into grocery stores. The *Enquirer* and *TV Guide* are among the top ten products in the grocery store

that make money. . . That's important information with which we can make counter-information like getting on public access cable to fight the ideological war that's going on here.

. . . What I want to talk about as a New York intellectual—who's an avid reader of the *National Enquirer*, although everybody in Queens apparently is completely embarrassed to read the *National Enquirer*; no one reads the *Enquirer* on the subway—that truth is stranger than fiction. As we all watch advanced capitalism or you could say decaying capitalism going down the tubes, we find a certain rise in the irrational. And the *National Enquirer* is about irrationality. Which is why I love it so much.

. . . I feel a certain ambivalence about dumping on the *Enquirer* or attempting to claim that the *Enquirer* is bad for your health because as far as I'm concerned the *Enquirer* is an interesting magazine. . . I love it. My ambivalence is about what you might call loving your chains. The *Enquirer* is not really like the *New York Times*. That's a false comparison I think. The *Enquirer* is more like a movie. It's narrative.

What I want to talk about is desire, because my desire is very split. Perhaps I am one of the millions of madwomen who survived the '60s and '70s, and are now living into the '80s. It's not *them* who are watching the soaps, which is the dominant genre on TV and film, but it's *us*. I'm watching the soaps, I'm reading the *Enquirer*. I know that it's a liar and a piece of shit and a rag and everyone's embarrassed to read it and embarrassed to carry it around, but I love it. That dialectic between my desire for the *Enquirer* and my knowledge that my mother would not be caught dead with the *Enquirer* anywhere in her home is the subject of this show.

Sarcasm and special effects. Didactic entertainment. What isn't funny is the insane asylums full of women, stars, escapism. More easy words. Stars. To live through these strange intimacies with creatures of celluloid and newsprint, themselves produced by multi-nationals, just as they produce our own perverse desires, contradicting the real needs. Can we remember them? Profound emptiness. As the economy and the texture of everyday life fall apart, the war budget rises, worklessness rises—with them our rising desperation. History shows us the choices are clear. People will learn to act and live for themselves again or the fascists will do it for us.

BRIAN WINSTON READS *TV GUIDE*

Good evening! I'm going to interpret a reading in the *TV Guide*. There's a problem about that because it's not easily considered as a "read"—in the sense that the New York magazine or *People* is a "read", much less the notion of commentary. Actually, it's really a bit like commentary. I don't know whether you recall last year, but *TV Guide* was running pro-family advertisements like this in the local paper of record: "America is an Endangered Species." The connection between the endangerment of that species and the television is not made apparent to the public.

TV Guide is not primarily a "read" because it is a list. I should qualify that by saying that its possible to write a book of lists in our culture and make it into a best seller, but *TV Guide* is essentially a list. It's rather like reading the telephone book. The list is a guide to the television of abundance. In fact, *TV Guide*'s most important ideological function is that it actually pretends that there is a television of abundance.

American television is extremely confusing to foreigners such as myself. We think it's all different and it's only after you've been here a couple of decades that you actually realize that it's simply repeating itself almost all the time. In this town, the broadcast stations, three of them, are in essence repeat-stations. Then there's PBS, which unfortunately doesn't have enough money to repeat either its own productions or productions that are released from my native land, Britain. There are the three networks themselves, which use program production codes like in the soaps during the daytime that are so slow in the way they consume material that it is possible to see them as repeats too. And then there is prime time—seventy shows or so a week. They manage to be more or less new ones—"fresh" is perhaps too strong a word to use in connection with the prime time network schedule. There are more or less new shows for only three seasons of the year. On the basis of this profligate amount of production, we now have a whole profligate number of extra channels, including the one I'm talking to you on now. The pay services on the extra channels that you pay for have discovered an extremely advantageous device which is that it is possible to show *Superman*, as they did last year, 53 times. The networks haven't quite caught on to that.

So *TV Guide*'s first function is to boast of a notion of abundance. It's in the nature of the case that every week it sort of pretends it's all different, but in fact it is largely the same every week.

There's an interesting problem that the publishers of *TV Guide* have had to face, namely, that there is a proliferation of channels. *TV Guide* was already difficult enough to read when it was only 2, 3, 4, 5, 7, 8, 9, 11, 13, 21, 25, 31, 41, 47, 49, 50. Now, you'll notice in the last few weeks that they've started to run a significant number of the pay services down at the bottom of their prime time table. This is an interesting indication of where their heads are at. They run arts, such as the *Alpha Repertory Television Service* brought to you by ABC for a couple of hours every night. They run *Bravo,* a cultural service brought to you by programming services out on Long Island. Other channels include CBS, which in this connection means CBS cable, other cultural services, ESN, HBO and various other movie channels as well as the super station, WTVS from Atlanta.

There's an interesting omission from this list, which is *Bravo's* partner in crime, as it were, a soft-core porno channel called *Escapade* (which is in conjunction with *Playboy*). The interesting thing about *Escapade* is that it shares a transponder 22,300 miles above us with the National Christian Network, which doesn't broadcast at night. I assume that up in the heavens all is cleansed, but what's equally interesting about this is that poor old *Escapade* watchers still have to guess, as *TV Guide* does not list those movies.

I have always found *TV Guide* extremely difficult to read and I suspect that the addition of the prime time box is a sort of acknowledgement of that difficulty. It's astonishingly easy to list things. They do us the courtesy, which I think is the most valuable thing they do, of listing the week's movies and special events of one sort or another right in the front. But basically, there is an endless problem of actually finding your way about, knowing what's on at any given point. I can only assume that since this is the bèst selling magazine in the country, most people don't have my difficulties with it.

There is a third level of immediate editorial comment apparent in *TV Guide*. Let me see if I can find you one. Here's one: where they give you boxes, they actually highlight what they consider to be the most important of the weekly events—sports and other things like that. I often think that this must be the most difficult editorial decision they have to make on a week-to-week basis. In the current edition, last night for instance, they decided to box Loni Anderson and Arnold Schwartzenegger. Arnold Schwartzenegger is to acting as Robert McKenzie, *TV Guide's* TV critic, it to television criticism. They decided to box this re-run rather than *American Playhouse* (one of the programs that was playing against it). *American Playhouse* is an extremely significant series—it's an attempt by PBS to get out of the clutches of the Brits and do some programming themselves. Last night's was called *Private Contentment* and was written by the southern novelist Ronald Price. Fortunately, it was directed by a British director—not even a television one—primarily known for his theater work. But nevertheless, it is interesting that when you're faced with the problem of what to box you should choose to box the re-run of a pretty run-of-the-mill filmstar biography made-for-television movie rather than to box something more significant, like *American Playhouse*. But most of the time there isn't enough stuff on television to box anyway.

The next thing that is important editorially, is their covers. Covers come in two sorts. There are covers like this: A noted historian judges TV holocaust films. Or this one, the current one that you are all probably reading now: Golda Meir. A few weeks ago it featured President Reagan, who was to talk about leaks, boycotts, and media bias. Those are sort of tangential to the actual substance of television programs, but most of the covers (two-to-one) are really covers of stars. This week we have the stars of the *Dukes of Hazzard*. We've had the stars of the cast of *Chips* and the cast of *Love, Sidney* and the cast of *Three's Company, Happy Days*, Tom Brockaw, etc. There's a sense in which *TV Guide* covers reflect it's somewhat schizophrenic and dual nature, where on the one hand it needs to bolster the television industry that it feeds off of but on the other hand feels a need to endlessly criticize it. This schizophrenia built into *TV Guide's* coverage is the most interesting thing about it.

PAPER TIGER TELEVISION (a partial list of programs)

Herb Schiller Reads the New York Times (6 half-hour programs)
Murray Bookchin Reads Time
Ynestra King Reads Seventeen
Tom Weinberg Reads the Wall Street Journal
Brian Winston Reads TV Guide
Ann Marie Bultrago Reads Agents' Names
Tuli Kupferberg Reads Rolling Stone
Bill Tabb Reads US News and World Report
Joel Kovel Reads Covert Action
Teresa Costa Reads Biker Life Style
Martha Rosler Reads Vogue
Paper Tiger Disarmament Special
Joel Kovel Reads Psychology Today
Sol Yurick Reads the New Criterion
Stanley Diamond Reads Scholastic Magazine
Paper Tiger Live Feedback Show
Sheila Smith-Hobson Reads Newsweek
Karen Paulsel Reads Computer World
Stuart Ewen Reads the New York Post
Seraphina Bathrick Reads Working Woman
Conrad Lynn Reads Commentary
Brian Winston Reads TV News
Joan Braderman Reads the National Enquirer
Tuli Kuperferberg Reads Sports Illustrated
Harry Magdoff Reads Business Week
Myrna Bain Reads Ebony
Alex Cockburn Reads the Washington Post
Archie Singham Reads Foreign Policy
Muriel Dimen Reads Cosmopolitan
Max Schumann on News Anchors as Expressions of Hysteria
Marc Crispin Miller Reads Cigarette Ads
Jean Franco Reads Mexican Photo-Novellas
F U CN RD THS SBWY AD: Kathleen Hulser Takes the A Train

Notes

[1] This position is clearly delineated in a publication directed at federal legislators, "Communications Airwaves: The Private Sector Option," by Douglas W. Webbing," *Heritage Foundation Backgrounder*, No. 224 (1982). The author proposes that property rights be granted broadcast licensees. His program for more private control could easily be translated into cable TV terms and applied to other information technologies as well.
[2] For instance, in a lawsuit against the New York State Cable Commission, Comax Telcom Corporation contended that cable operators should be protected by the First Amendment, and that allocation of channels should be solely determined by the operators. The suit was eventually dropped by the plaintiff.
[3] "Public Service Versus the Market," by Nicholas Garnham, in *Screen*, Vol. 24, No. 1 (January/February 1983), p. 13.
[4] *Ibid*, pp. 13-14.
[5] The integration of new information systems into the market economy, and the economic, and political impact of these developments is analyzed in Herbert Schiller's *Who Knows? Information in the Age of the Fortune 500*, (Norwood, N.J.: Ablex Publishing, 1981). For a detailed examination of the growth in one area of cable technology, videotex, see: *Pushbutton Fantasies: Critical Perspectives on Videotex and Information Technology*, by Vincent Mosco (Norwood, N.J.: Ablex Publishing, 1982).
[6] For a more thorough breakdown of these and other statistics on the concentration of ownership in publishing, see: *Many Voices, One World: Communications and Society, Today and Tomorrow*, the report of the International Commission for the Study of Communications Problems (New York: Unipub, 1980), pp. 104-106.
[7] ". . . [newspaper] groups [in the U.S.] own more than 60 percent of the 1912 daily newspapers. This trend to growth of chains continues; in 1978, of the 53 daily newspapers that changed ownership, 47 went into groups." *Ibid.*
[8] These are recent statistics from *Cablevision* (Dec. 10, 1984), p. 64.
[9] Glass's article, "Cracks in the Tube: How Socialists Can Break into TV," appeared in *Against the Current*, Vol. 2, No. 1 (Winter 1982), pp. 38-45.
[10] See Pat Thomson's "Independents on Television," *Afterimage*, Vol. 11, Nos. 1 & 2 (Summer 1983).
[11] *Brecht on Theatre: The Development of an Aesthetic*, edited and translated by John Willett (New York: Hill and Wang, 1964), p. 143.
[12] *The Mind Managers*, by Herbert Schiller (Boston: Beacon Press, 1973); *Who Knows?*, op. cit.
[13] The published report is titled *Many Voices, One World*, op. cit.
[14] See, for example, "Third World vs. the Media," by Philip H. Power and Elie Abel, *New York Times Magazine* (Sept. 21, 1980).
[15] For a concise summary of this debate on international communications, see Sheila Smith-Hobson's "The New World Information Order," *Fuse*, Vol. 6, No. 6 (March/April 1983), pp. 362-65.
[16] Brecht, p. 139.
[17] The special effects for the *National Enquirer* program were done by Manual De Landa.
[18] *Ibid.*, p. 146.

My shack has two rooms; I use one.
The lamplight falls on my chair and table
and I fly into one of my own poems—
I can't tell you where—
as if I appeared where I am now,
in a wet field, snow falling.

POET AT LARGE: A CONVERSATION WITH ROBERT BLY

Bill Moyers

ROBERT BLY:
I live my life in growing orbits
Which move out over the things in the world. . .

I have wandered in a face, for hours,
Passing through dark fires. . .

And I am gone to the desert, to the parched places,
To the landscape of zeroes!

I can't tell if this joy
is from the body, or the soul, or a third place.

BILL MOYERS: Tonight, Robert Bly reads his poems and talks about his life. I'm Bill Moyers.

When I told Robert Bly that I would like to talk with him on his home ground, he replied, "Then meet me in the north country." Here, on the shore of Kabekona Lake in Minnesota, and at a farm four hours further south, Robert Bly has grounded himself. He was born fifty-two years ago on the Minnesota prairie, three generations after his family arrived in America—Norwegian Lutheran pioneer stock.

He left the family farm a maverick, bound for Harvard, where he studied poetry with Archibald MacLeish. He first found solitude in New York City, where he spent three years living alone in a small room, working occasionally as a clerk or a house painter, eating one meal a day at the automat, reading hours at a time in the public library.

In time, he came back to his father's farm. He could live frugally here, could think and write. His first book of poetry, *Silence in the Snowy Fields*, captured solitude in the Minnesota landscapes. By the time he won the National Book Award in 1968 for *The Light Around the Body*, Robert Bly was writing about the Vietnam War, contemporary values, and the unconscious.

There's a rhythm to his life now. Two weeks a month he spends with his four children near the family farm in southern Minnesota. Four or five days a month he supports himself, barely, by giving poetry readings at colleges and in community forums like Cooper Union in New York. The rest of his time is spent here among the lakes and pines of the north country, where he translates poets from abroad and writes most of his poetry. Robert Bly thinks the best poets finally come home.

MOYERS: Were you as creative in New York as you are here?

BLY: Well, my own opinion is that no one is creative in their twenties, because you're too neurotic, you're too confused. I tried to write in my twenties, but I didn't do much. But how about you when you were in Texas? The problem is that no one wants to go back to the place where they were born. You *want* to go to New York, right? You *want* to do something like that. So here, you're not in Texas.

MOYERS: I'm not writing poetry, either.

(Both laugh.)

MOYERS: Have you spent a lot of time running from that northern Protestant Lutheranism in which men are not supposed to be expressive, in which tears are not allowed?

BLY: I don't know if the word "running" would be right. Fighting would be something like that. I tried for ten years to write poetry using only that male Protestant patriarchal Western European side, and it was a hopeless failure. And all through my twenties the poems were nothing. And eventually I came to realize that you have to have, you know, a feminine side to yourself also, the feeling side develop. And so I'm still involved in that. In the last five or six years I've learned how to weep for the first time.

MOYERS: You've learned how to weep.

BLY: I've learned how to weep.

MOYERS: Over what?

BLY: Sometimes out of thinking of the suffering of my own life; instead of having a stiff upper lip, which was what you recommended, you weep. I remember weeping went on during the Vietnam War, over other people's suffering. And evidently weeping is something that's done a lot in the world. You wouldn't know it, to be in Minnesota, but evidently when people find something beautiful they weep. One day I was listening to a symphony, a Schönberg thing, and the snow was falling and I started to weep, I started to weep for an hour, I couldn't stop.

MOYERS: One of the best lines that appealed to me is that line where you talk about the parts of us which grow when we're far from the centers of ambition.

BLY: Far from the centers of ambition, right. So, that's the kind of gift given you if you move back to your crummy little place. At least you're far from New York or Los Angeles. And I understood that best when I'd been moved out to the farm after being in New York a while, we moved out to the farm, we had a horse; we didn't have any money, so I didn't have any water in the house, so we had an old windmill. So I went out to water my horse one day, and I'd been reading the Buddhists. And I was told in college that I was supposed to be ambitious as a poet. Send my poems out, get them published, get them into *New Yorker*, whatever—right? Buddhists say no, that's all wrong, you don't understand, that's completely wrong. If you want to be a poet, you have to merge with the trees and the animals and so on. Now, those things don't have any ambition. So therefore you *can* merge with them. So therefore they try to teach not-ambition. So I wrote a little poem,

How marvelous to think of giving up all ambition!
Suddenly I see with such clear eyes
The white flake of snow
That has just fallen in the horse's mane!

(At Cooper Union.)
BLY: "Six Winter Privacy Poems." Part one.

About four, a few flakes.
I empty the teapot out in the snow,
feeling shoots of joy in the new cold.
By nightfall, wind,
the curtains on the south sway softly.

Part two.

My shack has two rooms; I use one.
The lamplight falls on my chair and table
and I fly into one of my own poems—

I can't tell you where—
as if I appeared where I am now,
in a wet field, snow falling.

Third poem.
More of the fathers are dying each day.
It's time for the sons and the daughters.
Bits of darkness are gathering around them.
And the bits of darkness appear as flakes of light.

A little meditation poem.

There is a solitude like black mud!
Sitting in this darkness singing,
I can't tell if this joy
is from the body, or the soul, or a third place.

(Scattered laughter, applause from audience.)

BLY: (Laughs.) A little poem, "Listening to Bach."

Inside this music there is someone
who is not well described by the names
of Jesus or Jehovah, or the Lord of Hosts!

Here's a last little one, the sixth one.

When I awoke, new snow had fallen.
I am alone, yet someone else is with me,
drinking coffee, looking out at the snow.

(Applause from audience.)

(In Minnesota.)
MOYERS: When did you know for sure that you wanted to write poetry for a living?

BLY: It was in college, and I was studying Yeats.

MOYERS: At Harvard.

BLY: Yeah; and we started to work on Yeats, and I was astounded how much could be contained inside a single poem. Of the man's life, and of his opinions, and of the past, and of his occult work and of his own private thought. I had always thought of poetry as sort of like a bird singing something eccentric that happened at the edge of something. And suddenly, reading Yeats, you realize, as I've realized since, that the entire work in psychology—if you go into poetry and write it, you're drawn to the most amazing things that have happened in the last hundred years, which is psychology, the advances in psychology. You're also drawn into politics and you're drawn into the whole question of nature, and you're drawn into what is the collective—you follow me?

MOYERS: Yeah.

BLY: It was when I realized that inside a poem you could find the entire world and work in it.

MOYERS: So when someone asks you, From where does a poem come, your answer is, Everywhere.

BLY: Yes, everywhere. But I suppose especially from, let's say, the dark side, or something. Since human beings became civilized, they have a tremendous liking for the light side of themselves, the side on which the light falls—the civilized side, the polite side, the economic side. But actually underneath all of that there's a dark side, which is sometimes referred to as a shadow. So therefore in the shadow is contained everything in yourself that you don't like, as well as your whole primitive self. So if someone says, Where does the poem come from, in most cases it comes up from the unnoticed side of every human being, it comes from the primitive part of him.

MOYERS: I read someone who said, "Robert Bly's poems are a journey into the interior. He's down there somewhere, waiting for us."

BLY: (Laughing.) Maybe, if I haven't left. But I like that idea very much. In myself, what I have been trying to do, I suppose, in the last—well, I'll read you a poem which I wrote in New York; I'd been living in New York and then I left and went down to—I was heading for Charleston, I went across the Chesapeake Bay. And I hadn't seen water for two or three years. And moreover, I felt while I was living in New York that I would probably never do anything, that my poems would probably be a failure, that I would never publish a book of poems. And yet I received enough nourishment from simply writing and reading poems that that was all right with me. So this is mentioned in the poem; I'll read it to you. "On the Ferry Across Chesapeake Bay."

On the orchard of the sea, far out are whitecaps,
Water that answers questions no one has asked,
Silent speakers of the grave's rejoinders;
Having accomplished nothing, I am traveling somewhere else;
O deep green sea, it's not for you
That this smoking body ploughs toward death;
And it's not for the strange blossoms of the sea
That I drag my thin legs over the Chesapeake Bay;
Though perhaps by your motions the body heals;
For though on its road the body cannot march
With golden trumpets—it must march—
And the sea gives up its answer as it falls into itself.

MOYERS: You're fortunate to catch moments like that and hold them forever. What you call peak experiences.

BLY: Yeah. And I was glad to do that and feel it, to see it actually in the water. And usually that experience doesn't come until you're twenty-nine, maybe, or thirty—maybe in your thirties—when finally you forget what the collective wants. Forget it. And forget what your parents want. And you slowly fall into yourself.

MOYERS: You're off on your own.

BLY: And that's a wonderful moment.

MOYERS: You said at the reading, or in one of your poems, that you aren't sure whether a man's feeling of joy comes from his body, or his soul, or a third place. What is that third place?

BLY: Well, I was doing some meditation during that poem. It was a little poem that goes,

There is a solitude like black mud!
Sitting in this darkness singing,
I can't tell if this joy
is from the body, or the soul, or a third place.

And I suppose there's a little joke in there, because we're told we're either body or soul; there must be more possibilities than that. But what happens, I think—I'll just go back a little. We're at about the seventieth or eightieth anniversary now of Freud's first work in bringing the unconscious forward. And Freud said, you know, you have a conscious and light mind, and underneath that you have a dark, unknown mind. And Christianity did a great deal of harm in that it called the unconscious the devil; therefore it blocked access to that mind for centuries. So it's a very important thing that has happened when Freud and Jung, eighty years ago, and Adler and others began to give attention to the unconscious mind. That's exactly opposite than calling it Satan. Actually, what Freud did is say you have two minds and the dark one may be the more intelligent of the two. The dark one is the mind that makes up your dreams. And Jung later showed how incredibly intricate and complicated and spiritual, actually, the dreams are. So this is my idea: about twenty years ago an important change occurred in the psyche—or at least of Western man—and that is that after human beings in the West, through Freud and Jung, started to pay attention to the unconscious in a positive way, to regard it as something positive, it began to get more confidence, just as if you pay attention to a child it'll start to blossom. So the unconscious has now responded by creating something which was not here before. It's as if the unconscious is now objectified and is out there somewhere. Now, I don't know, I don't understand it very well, but it's something like this: for a thousand years now there has only been in the West man against nature. Man is intelligent as Descartes says, and wonderful, and nature is stupid and doesn't think and is full of snakes and alligators and so on; a naked confrontation between man and nature. What I think has happened is that the attention to the unconscious has created a third being, which now stands between man and nature, and that's a wonderfully joyous occurrence.

MOYERS: But which joins them, doesn't it?

BLY: It joins them. It can act as something between man and nature, as if the old unconscious of human beings now, which has been for thousands of years, millions of years, has agreed to come out and take a stand between man and nature. So there isn't that naked split anymore.

MOYERS: And this is that third place.

BLY: I think so. I don't know what to call this. I called it like the third being which is now present, and one of the reasons that there's so much excitement in poetry now is because in the old days, let's say, let's say in the eighteenth century, either Pope wrote a poem about man—the proper study of mankind is man; and he stayed completely with the man—or Keats would write a nature poem. And it was as if those were the only two choices. But now there's this third presence inside; and what is exciting about poetry since the 1920s is that sometimes the poem goes into this third being and comes directly out of it. And it's neither man nor nature; I can't explain it any more.

MOYERS: How do you draw it out? It's not like putting a bucket down into a well.

BLY: I think that, of course, everyone who wants to get near it, in a way, would follow the road that Freud and Jung took. In other words, you pay attention to your dreams. If you have something by your bed, you write down your dreams. Because the dreamer is entirely a part of this third place, or it's entirely a part of the unconscious. And so you pay attention to nature a lot. All Chinese poetry, which had a lot of third presence in it in ancient times, paid terrific attention to nature, because this being is close to nature, as well as to yourself.

MOYERS: Is this how you write your poems?

BLY: When I started to write, I was writing completely with my conscious mind. That's the way I was taught to write. And I wrote for ten years without any success. And by being alone, I began to sink down a little bit into the unconscious, by being alone a long time. And now I think of it differently, a little bit differently now. I'd say that there's that one moment in which you fall into yourself, hmm? And that's a moment in which you feel the unconscious going to hold you up; it's going to hold you up. It isn't necessary to be successful, you're *already* successful, you understand me?

MOYERS: It's like swimming in the dead sea.

BLY: Yes! You feel a support underneath you, made of your own feelings, and so on. And that's a wonderful moment.

MOYERS: But most of us don't trust our feelings that much.

BLY: No, and that's a great problem of being a male in the West. We're just not taught to do that. So you could also say that you're being held up by the ocean. After all, the name of the ocean is *mare*, which is the same word as Mary. So in a way, the ocean also stands for all the feelings and all the feminine parts inside a man, which one learns to trust and to help hold him up.

(At Cooper Union.)
BLY: So I'll give you a poem about a couple of my children. I have two

daughters, one's sixteen, one's fifteen, called Mary and Biddy, and this is a poem which touches on them. We don't have a television set, never had, so when we get snowed in or something, then the girls—and I have a couple of little boys, too—what happens is that they usually make up a play or something, and my daughter Mary usually casts herself as the princess, and the boys get the crummy parts, because they're littler. . .

(Laughter from audience.)

BLY: And then they cast it and put on the costumes and put it on, and Sunday night the grownups are invited up to watch it for a small fee.

(Laughter.)

BLY: And so this describes one of those nights.

Ah, it's lovely to follow paths in the snow made by human feet. The paths wind gaily around the ends of snowdrifts, they rise and fall. How amazed I am, after working hard in the afternoon, that when I sit down at the table, with my elbows touching the elbows of my children, so much love flows out and around in circles. . .

Each child flares up as a small fire in the woods. . . Biddy chortles over her new hair, curled for the first time last night, over her new joke song.

Yankee Doodle went to town,
a-riding on a turtle,
turned the corner just in time
to see a lady's girdle. . . (that's a Taurus)

Mary knows the inscription she wants on her coffin if she dies young. . .

(Laughter from audience.)

. . . and she says it:

Where the bee sucks there suck I
In a cowslip's bell I lie. . . (couple of lines from Shakespeare)

She is obstinate and light at the same time, a heron who flies pulling long legs behind, or balances unsteadily on a stump, aware of all the small birds at the edge of the forest, where it's shadowy. . . longing to capture the horse with only one hair from its mane. . .

And Biddy can pick herself up and run over the muddy river bottom without sinking in; she already knows all about holding, and she kisses each grownup carefully before going to bed; at the table she faces you laughing, bent over slightly towards you, like a tree bent in wind, protective of this old shed she is leaning over. . .

And all the books around on the walls are light as feathers in a great feather bed, they weigh hardly anything! Only the encyclopedias, left lying on the floor by the chair, contain the heaviness of the three-million-year-old life of the oyster-shell breakers; oh, they were so long, those dusks—ten thousand years long—where they fell over the valley from the cave mouth (where we sit). . . And the last man killed by flu who knew how to weave a pot of river clay in the way that the wasps do. . . Now he's dead and only the wasps know. And the marmoset curls its toes around the slippery branch, aware of the furry chest of its mother, long since sunken into a hole that appeared in the afternoon. . .

Well, dinner's finished, and the children pass out invitations composed with felt pens.

You are invited to

"The Thwarting of Captain Alphonse"

Princess Gardner: Mary Bly
Captain Alphonse: Wesley Ray
Aunt August: Biddy Bly
Railway Track: Noah Bly
Train: Sam Ray

Costumes and Sets by Mary Bly and Wesley Ray

Free Will Offering Accepted

(Laughter, applause from audience.)

BLY: I'm going to give you a poem now, I'll give you a poem of my own which is connected a little bit with grief. And it's a poem sort of out of the '70s. And it's called "Snowbanks North of the House," and it's about those things that happen in Minnesota when snow comes down from the north and comes from Canada or somewhere, and it's everywhere except right around the house. North side of the house there'll be a gap of five to six feet, and I'd noticed that since I was a boy. And finally I wrote this little poem about it.

Those great sweeps of snow that stop suddenly six feet from the house,
Thoughts that go so far.
The boy gets out of high school and reads no more books.
The son stops calling home.
The mother puts down her rolling pin and makes no more bread.
And the wife looks at her husband one night at a party,
And loves him no more.

And the energy leaves the wine,
And the minister falls leaving the church.
It will not come closer.
The one inside moves back,
And the hands touch nothing, and are safe.

And the father grieves for his son, and will not leave the room—(this is Lincoln)
And the father grieves for his son, and will leave the room where the coffin stands;
He turns away from his wife, and she sleeps alone.

And the sea lifts and falls all night,
And the moon goes on through the unattached heavens alone.
And the toe of the shoe pivots in the dust,
And the man in the black coat turns and goes back down the hill.
No one knows why he came,
Or why he turned away and did not climb the hill.

(Applause.)

BLY: Well, that has a little soul in it, you feel a little soul in it? Because finally I myself am willing to accept a little grief.

(In Minnesota.)
MOYERS: I want to be sure I understand what you mean when you're talking about Hades, grief and pain. Are you saying that in order to become human, to discover that third place, we have to admit, acknowledge the presence in our lives of evil?

BLY: I don't think it's evil; it's connected with the pain and the grief we have in our own background, and as you know, in America especially, the men and the women are taught to be cheerful all the time. And pain is considered something bad that you're ashamed of. And if you go to a Kiwanis meeting you hear them insanely optimistic about everything. So therefore there's a movement in the other direction, in which you move downward towards grief and acceptance of the side of yourself that you don't approve of, that your parents didn't approve of. And in alchemy this is described as lead. You know, they say, Don't begin with the spirit, that's the wrong place to begin. The problem with much charismatic Christianity is that of going too fast into the spirit; they're going into the light. Alchemy says no! the spiritual process begins with lead. It begins with the element lead, which is a part in you that's so heavy and so deep-lasting you don't think it'll ever change. And so lead is connected with the grief in your family, it's connected with your alcoholic grandfathers and your insane grandmothers and the suicides in your family, the part in your family that always seems to get blocked, generation after generation. That's called lead. And you go down there and begin with lead and live in lead, and you just remain down there for a while; that was the idea of alchemy. And then, also, the lead involves the grief of your own country. So therefore an alchemist would say, Well, if you really want to be a poet, you know, you're going to have to read a lot about American history, because a lot of your lead is contained in all the suffering and grief that America has caused. Well, do you understand me?

MOYERS: But you're not saying that we ought to grovel in it.

BLY: No, no. There's the opposite point of view, many people are becoming addicts now, addicts of anger; it's like being an addict of heroin. That's not the same process. The anger is taking them over. But the image of going into Hades, you go into grief, down through Hades, because you're going somewhere else. That whole process of slowly the making of the soul. You don't stop and become an addict of grief or an addict of anger.

MOYERS: You move on.

BLY: Yeah, the point is to—in ancient things, you know, in Eleusinian mysteries they give you like a twenty-five-year process, and they slowly feed you through. And again, unconsciousness is not the same thing as grief. Unconsciousness, being unconscious, is connected with rage. Okay, you're a father. One of the things I noticed is something like this: if you're unconscious a lot, as I have been for many years, if you're unconscious everything seems to go all right except you feel a lot of anger and rage. And that's all right until you become a father. When you become a father you may notice that your rage will suddenly fly out at a child. And that child has done nothing to deserve that rage. Brohhhmm! it goes, and you feel it leave you. And so therefore if you're a father you realize that one should do something about that; and of course if the rage is connected with being unconscious, then the only thing to do is to try to become more conscious. So I made a little progress there, working on that, especially in relation to a son of mine named Noah, who's about twelve now. And I'll give you a poem I wrote for him. It's a new one.

Night and day arrive—day after day goes—
Night and day arrive, and day after day goes by,
And what is young remains young and grows old,
And the timber pile does not grow younger,
  nor the two-by-fours lose their darkness.
But the old tree stands so long,
The barn stands so long without help.
The advocate of darkness and night is not lost.
And the horse steps up, turns on one leg, swings,
And the chicken, flapping, claws up onto the roost,
It wings whelping and wholloping.
But what is primitive is not to be shot out
  in the night and the dark.
And slowly the kind man comes closer,
  loses his rage, sits down at table.
So I am proud only of those days that we pass
  in undivided tenderness,
When you sit coloring a book stapled with
  messages to the world,
Or coloring a drawing of a man with fire
  coming out of his hair,
Or we sit at a table, with small tea
  carefully poured.
So we pass our time together, calm
  and delighted.

MOYERS: And you're saying to your son. . . ?

BLY: Yeah, somehow the more conscious we become, which I feel is connected with eating our own shadow and eating our own grief, then the more we are able to be calm and quiet with our children, and be close to our children.

MOYERS: What about male-female relationships? You've been thinking a lot about those. At the reading the other night I heard you say that the pain men and women give each other is exactly what they need.

BLY: (Laughing.) Well, it's sort of a joke, but I mean, America has so much comfort in it now, and you know, everything is centrally heated and cars have such wonderful heaters in them; and so where are you going to find pain except in the relation between men and women? (Laughs.) So it's the only suffering left to us.

MOYERS: But you talk about developing the female side of our natures. When did you decide that's what you had to do? Was it with marriage?

BLY: Before I was married I was aware, living alone in New York, I was aware of how barren the personality is when only one half is developed. And I wrote a poem called "A Man Writes to a Part of Himself," which gives that feeling in the twenties when you go back to your room and you're twenty-six years old and the room is so barren and so lonely you want to leave right away. And I realized that the feminine side of myself was living off somewhere. Actually, in a poem I put her in a valley, and actually she had regressed and was like living in the cave times. Probably that's true. My male side was twenty-six and living in a city, my female side was living 2,000 years ago; and the problem is to bring those two together, somehow.

MOYERS: How do you do that?

BLY: Well, again, the process of paying attention to dreams is helpful—I don't know, I find, as a male, a lot of my optimism in my male side and lot of my grief and feeling in my female side. So therefore one could say it has to do with feeling. It's what Robert McNamara did not do. When you said to him, You know, they're burning children in Vietnam, do you know that? They're burning them with napalm. And McNamara, he'd answer with his male side completely. He'd say, Okay, I'm pulling down the charts now. I want you to see these charts. Now we're bound to win.

And that's not an answer to the question, because he's not answering with his feeling side at all.

(At Cooper Union.)
BLY: I was saying, you know, one thing the American can do is just to become the collective; that's a great solution. And especially you want to become the male collective. So I'll give you a mask of a man who has done no development on the female side at all. It's a poem I wrote a long time ago called "The Busy Man Speaks," and I'll give it with a mask. (Puts on mask.)

(Laughter, applause from audience.)

BLY:

Not to the mother of solitude will I give
myself away!
Not to the mother of art, nor the mother
of conversation!
Not to the mother of tears, nor the mother
of the suffering of death!
Not to the mother of the night full of crickets,
nor the mother of the open fields!
Nor the mother of Christ!

But I will give myself to the father of righteousness,
who is also the father of cheerfulness,
Who is also the father of perfect gestures!
From the Chase National Bank an arm,
a flame has come,
And I am gone to the desert, to the
parched places,
To the landscape of zeroes!
And I shall give myself to the father of righteousness,
the stones of cheerfulness, the steel of money,
The father of rocks.

(Walks around audience.)

I'm going to get you. You think 'cause you go to Cooper Union I won't get you? I'll have you within five years. You know what I'm called? The Industrial Revolution! Every year since the Industrial Revolution I'm stronger! You think you can fool with machines and you won't look like me? Every time you drive your car you look more like *me* when you get out of it.

(Laughter, applause.)

BLY: Want some labor-saving devices? Half the women in the United States look like *me* now! I've got your parents already. Go ahead, laugh, you'll laugh at your funeral, won't you? I'm not so easy to be gotten around. Television is my specialty.

(Applause.)

BLY: Including Channel 13.

(Laughter, applause.)

BLY: You know what I like about television? 'Cause you can be passive watching it. You don't have to do *anything*. The programs are all *educational*. I love passive students. I can eat 'em. Eat em! Go ahead. More rock records, more *me*! Get yourself an instrument, I'm not so sure about that. But as long as you've got your records, hrrnnhh!

(Applause.)

(In Minnesota.)
MOYERS: Darkness comes early to your Minnesota winter.

BLY: Hmm. . . yes. The pines call it down. You want a poem about that? (plays dulcimer)
MOYERS: Sure.

BLY:
There is unknown dust that is near us,
Waves breaking on shores just over the hill,
Trees full of birds that we have never seen,
Nets drawn down with dark fish.

The evening arrives; we look up and it is there,
It has come through the nets of the stars,
Through the tissues of the grass,
Walking quietly over the asylums of the waters.

The day shall never end, we think:
We have hair that seems born for the daylight;
But, at last, the quiet waters of the night will rise,
And our skin shall see far off, as it does under water.
And our skin shall see far off, as it does under water.

I don't know if the dulcimer's any good with that or not.

MOYERS: But you like the darkness, don't you? Not exclusively, but naturally.
BLY: It's the other half. I found almost all of the poems I had written in *Silence in the Snowy Fields* were written at dusk, at the passage from day to night, and that's possibly a time when the unconscious opens.

MOYERS: Do you think in the end the light persuades the darkness?

BLY: I don't know; I don't know. I read St. John, but I'm not sure I believe it.

(Both laugh.)

MOYERS: I know you've been doing a lot of work in fairy tales for modern men. Why?

BLY: You know, in the last ten years there's been talk about women's growth and stages of women's growth, which has been very helpful to everybody. But less talk about the stages of men's growth. And it's possible that the fairy tales were composed, some of them, ten to twenty thousand years ago by men and women of the nature of Jung and Freud and Karen Horney who found themselves not only having amazing insights on growth processes in human beings but they found the necessity to put it—because libraries were being burned they couldn't simply print it—so they had to put it into a story, extremely vivid, that would be remembered for a long time. And one example—well, I thought I might begin the book with "Jack and the Beanstalk."

"Jack and the Beanstalk" says that in the beginning the mother tells the boy to sell the cow for money, and he starts out in this task. But he disobeys his mother. So the first movement of the boy, when the growth begins, the first act is to disobey the mother. And strangely, in this poem the mother in a way represents the system; she asks him to buy into the system by selling the cow. And that was all going well, except that he met an old man—a wise old man. Which brings in the whole area of magic and respect for intelligence in old men and things like that, and the old man offers him instead three magic beans. So the word "magic" suddenly appears. And so the second stage in the growth of a male is when he approaches the world of magic; and certain boys are open to magic still.

MOYERS: His first stage is disobeying his mother.

BLY: Yeah—apparently; according to this story.

MOYERS: His second stage is accepting magic.

BLY: Yes. Now, the growth can stop at any point. So therefore in this case the boy is open to magic; don't know why. Maybe his parents read him fairy stories. Don't know why. But he's open to magic, and he accepts the three magic beans, okay. Then he comes home with the three magic beans, right? His mother says, "What are you doing? Lookit, you moron! You've bought these three beans—what is this?" And she's furious and throws them out the window. But that helped the beans to touch earth. And the next day the beans have grown up, and there's this large green thing outside the window, which, a Freudian might say it's his penis, but we hope not because later it's cut down. Anyway, it's his manhood, perhaps. His manhood is now growing up, with the help of the old man, and with the help of the anger of his mother. If the mother had not thrown the beans out, if she'd said, "Well, how wonderful. We have a couple of beans now. You and I'll just talk about it, we'll put them in the cupboard, we'll have something to talk about." It's important the woman go through the anger, too. So she throws them out. So then he has three choices: one is to ignore it, close the curtain, live with his mother; second choice is to cut it down, which was recommended in the '60s—many males were recommended to get rid of their aggression and their anger, cut it down; the third choice he has is to climb it.

Jack decides to climb it. So he decides to climb his own manhood, hm? So it turns out that if you start to climb your own manhood, when you get up on top, you know what you find? You find a *really* big male. And this is related to finding your own father, who looks down and says, "What is this? My son is growing up? He's going to be a male? Arrghhh!" In that area he meets his father; in another area what he just did was to meet a collective like IBM. Like American Tel & Tel, who are enormous giants, who specialize in eating young males. And a third possibility is that he's meeting some large male on some other plane; you know, on some spiritual plane he's meeting a very large male. In any case, it's a dangerous business. And what this story says that's so wonderful is that the male will not survive his contest with IBM unless he has a feminine presence to help him. Inside the castle there's a young girl at a spinning wheel. Nobody knows what this is, very mysterious. But when the male comes she hides him in the oven. Always hides him. There are many versions of "Jack and the Beanstalk," but always she hides him in a sewing basket, in an oven, in some feminine area. So he's protected by the feminine twice, first by the woman who protects him and second by being in the kitchen area. And then the giant can't get him. Then an interesting thing happens. Because he's gone through the confrontation with the father but he lived—with the great father, with the collective father—only with the help of a woman. . . or maybe his own feminine side. Then, if the male does that, he receives energy. In the fairy story that's described—oftentimes in fairy stories when they're talking about energy, increase of energy, they'll use gold coins. He steals gold coins from the old father, and then runs down the tree, and he has the gold coins; but shows how difficult it is for the male to get away from the mother because as you know, what happens, he and his mother spend the coins. And then he has to go back up the tree and make the confrontation a second time.

This time he gets a goose who lays golden eggs. So the energy's becoming a little more organic now, hm? More tied in to his body. It's a mammal that's laying the eggs now. So he again succeeds with the help of the woman, goes down the tree, but again he can't get away from his mother. He and his mother actually get impatient and they kill the goose in order to get the golden egg. Ah, it's touching how hard it is to get away from his mother. And so then he goes up the third time, and you remember the third time he brings down a golden lyre, and now it's much more integrated in the whole world of music and sound and all of those things. Now it's a self-containing energy that continues. The goose is sort of the level of Balzac, who had a lot of energy, but the lyre is in the area of Kabir and the great religious singers who have endless supplies of energy. And then he comes down the ladder, and this time the lyre calls out and the great collective or the great father follows him down the ladder and he cuts it off and the father is killed. Then he's free of his mother, too.

MOYERS: If you get us men to reading fairy tales, you may destroy the economic system.

BLY: Well, I'm afraid it'll never happen. What did Isaac Bashevis Singer say? You can't affect the system, you can't even make it worse.

(Both laugh.)

MOYERS: Why do you want us to read fairy tales? You think it does help us gain insight into the meaning of myth?

BLY: I think that in ancient life, for example, they knew much more about the stages of growth than we do. The whole idea of being saved instantly has been a great destructive power in understanding that our growth proceeds by stages. Yeats said it proceeds by spirals. And we mustn't expect to grow up in our twenties; that's a disaster. We ask great creativity and great spirituality out of our people in our twenties; no, that's not what they are. Their job is to make a connection with the world and try to do something in the world, and then, as they get older, it'll be time for spirituality or for growth. It's important to do them in the right stages, and fairy tales are unbelievably succinct and daring in the way they lay out certain stages. And most fairy stories end with the prince and the princess being married. That has nothing to do with marriage. The fairy stories proceed to the point at which the male inside the body and female are married. That is the marriage that takes place in the castle. And at that point, then, you're prepared to do something in life. When the male and female have been married, when the king and the queen have been married.

MOYERS: When you are whole.

BLY: When you're whole, hopefully—even for a minute or two. (Laughs.) You'll probably fall back later. (Laughing.) (Playing dulcimer.) So that's why it's exciting. There are these most wonderful Eskimo stories, and ancient stories about man's growth.

MOYERS: Where are you, in what stage?

BLY: Don't ask me, don't ask me. (Laughs.) Probably at the bottom. Probably haven't even climbed the beanstalk yet.

MOYERS: Move over, that makes two of us.

(Both laugh.)

MOYERS: You were very popular with college students in the '60s. I know because I heard from them, I saw them, I met them, and part of it had to do with your very early opposition to the war in Vietnam. Was it just the war that offended you, or did you see something else more deeply in the American psyche?

BLY: Well, I had already—I had published *Silence in the Snowy Fields*, and I was thought of as a poet who was interested in trees and snow and stuff. But I'd also written another book, which no one would publish; it was called *Poems for the Ascension of J.P. Morgan*. And no one wanted political poems at that time, and no one would print it. And what I did was to take a poem about our whole capitalist general situation and then put next to it an advertisement. So it would alternate between a poem and an advertisement; you know, insane things out of *Time* magazine, and stuff like this. So I was interested in the whole issue of what's it like when a nation tries to suppress and pretend that it hasn't done the murder of the Indians. And it was out manifest destiny, you know: God *asked* us to go to the West Coast and wipe out all the Indians, you know, really, he did. And when you read that stuff in history books, you realize that we're engaged in a vast forgetting mechanism. And from the point of view of psychology, we're refusing to eat our grief, refusing to eat our dark side, we won't absorb it. And therefore what Jung says is really terrifying—if you do not absorb, you know, the things that you have done in your life, like the murder of the Indians and bringing the blacks in—if you don't do that, then you will have to repeat it, you continue to repeat it. There's a repetition compulsion which neurotics have. And America is neurotic. It has the compulsion to repeat the Indian wars. So as soon as we started to go into Vietnam it was perfectly clear to me that what was about to happen was that the generals were going to fight the Indian war over again, and the Indians—I mean the Vietnamese—are poor in relation to us, they're Asian, somehow they have this strange religion called Buddhism, which is not Christian; the best thing to do is just to kill them. The best thing to do is to get a good excuse and wipe them out. And so I felt it couldn't help the United States at all because all it would mean is we'd go deeper in—deeper *into* neurosis. And you give Eisenhower credit; Eisenhower saw that something like that was going to happen. He refused to send troops to Vietnam. That was the greatest thing he did as a President. And he said, No, what can we do over there? It's hopeless. And he said no to the French. And then Kennedy, who was an idiot in many ways, just a wonderful, intelligent idiot, didn't see that at all. He was so crazy about the light side of life, he was so crazy about elegance and good humor and all of that, and he didn't see what he led America in for. I blame Kennedy most, and after that, Johnson.

MOYERS: That was a very bitter time in your life, it seemed to some people. You talked about the time when the ministers lie, the professors lie, the priests lie, television lies.

BLY: All of this is a longing for death.

MOYERS: A longing for death? You think our society has a kind of death wish?

BLY: Yes, that if you don't eat grief and you don't face out what happens, it's so strange; what happens is that the death longing, which is just as strong as the life longing in us—it's right there. As Freud at the end of his life said, I've overlooked something; I said that eros ruled everything. I was wrong about that. There's eros and also the death instinct—so the death instinct than can become powerful in a human being. If it becomes more powerful than the life instinct, the death instinct will come forward and he'll commit suicide. If the death instinct becomes strong in a nation, as it was in us around 1960 and '63 and '64, if the death instinct really becomes more powerful than the life instinct, then it will do something absolutely destructive to itself as a nation.

MOYERS: And how do you get it open? How do you get it out? How do you deal with it?

BLY: I don't know. I don't know. I think America will do it again; you know? I think the next war will probably be the Brazilian one. Because we've learned *something* from the Vietnam War; but as I say, I miss the grief. I don't see any real grief in the United States over that. The people that are suffering the most are the veterans who went and did what we asked them to do. Then they come back with their leg blown off and no one'll even talk to them. You know, people come up to veterans and say, "You deserved it! You got what you deserved!" Well, that just shows we aren't facing it, we still aren't seeing what we did.

MOYERS: What legacy do you think the '60s and Vietnam left us?

BLY: It left a wonderful legacy of the knowledge that we're quite insane.

MOYERS: Insane?

BLY: Quite insane, America's quite insane. And that was the gist of the whole thing; I mean, we had our best people, McNamara, Johnson, everybody in there, and they couldn't figure out the most elementary thing, namely that they were going to lose. You can't fight against a people who are actually fighting for independence and their own honor. So the fact is that we had our best brains working at it and it didn't work, so insanity's the only other possibility. I mean, America has a profound insanity, and it's all right. France is insane; I mean, why shouldn't *we* be insane?

MOYERS: What is insanity? Denial?

BLY: I don't know. I don't know what insanity is, but I would say that if you ignore your dark side long enough—a man can ignore his dark side for ten to fifteen years, and then it begins to get strong out there itself. Everything you ignore in your personality is pushed out to the edge, where it begins to gather strength and begins to be hostile to you. And eventually it'll come back in and bust up your whole—you know, serene ...life. And so I think that insanity is when the dark side has been ignored for years in a puritanical society, and then suddenly it returns and smashes everything. And what does it cause us? A depression. It causes us a depression, which is a wonderful thing. The unconscious knows that our invasion by the insane side produced B-52s taking off from Guam. What did they cost? $50,000 every time one took off? Something like thirty of them took off every day for five years; the result is an economic depression. But the word depression is wonderful, because it's chosen by the unconscious to make you understand that the economic world and the psychic one are identical.

MOYERS: So what does poetry have to say to this kind of society?

BLY: I don't know—it says, it says—I don't know what it says. But I notice the wisdom of people like the Hopis, for example, who will have two to three days a year, or even a season, of mask work, when everyone goes out and is insane, you know? They're totally insane. (Laughs.) All that's wonderful; everybody's a little insane, and then in the end they're all sane from agreeing to be insane. Whereas you go to a Lutheran church, you notice anybody insane in a Lutheran church? Baptist church?

MOYERS: Not in a Baptist church. Lutherans maybe, but not. . .

(Both laugh.)

MOYERS: Is that why you use masks?

BLY: Yes, I use them for that purpose.

MOYERS: Should we have a kind of national mask day?

BLY: Yeah, that'd be helpful, um-hm. That'd be helpful. Everyone agree that they're partially insane; it would take a lot of load off the people in the insane asylums, who have to carry the whole weight of insanity. (Laughs.)

MOYERS: There's a poem—one of the first poems you published, in *Light Around the Body*, dealt with some of these themes. I remember it was one of the first poems of yours I read in the '60s. "Come With Me." Remember that?

BLY: Yes, I remember that, and I thought of it a minute ago when we were talking about how around 1960 or '63 or '64 it began to be apparent that we were going into extremely dark areas without willing it. And I use images here of an old tire rolling down a hill, and then drowning in some slew at the bottom. And to me what happened then was that this darkness began to grow, and then Kennedy was standing in the way; because Kennedy had too much to do with light to a certain extent. And I remember the day that Kennedy was killed; it was perfectly apparent to me—I got this horrible feeling, Here we go; here we go; now we're going right down into the dark now, going right down. And you know, it's as if the unconscious got Kennedy out of the way because—I mean, he was foolish, but he had wonderful connections with light and things, and if the country's going to suffer, you're going to have to get rid of a man like that. So it can suffer. (Plays dulcimer.) So the poem is called—

Come with me into those things that have felt this
despair for so long—
Those removed Chevrolet wheels that howl with a
terrible loneliness,
Lying on their backs in the cindery dirt, like men
drunk, and naked,
Staggering off down a hill to drown at last in a
pond.
Or those shredded inner tubes abandoned on the
shoulders of thruways,
Black and collapsed bodies, who tried and burst,
And were left behind;
Or those curly steel shavings, scattered around
on garage benches,
Sometimes still warm, gritty when we hold them,
Who have given up, and blame everything on the
government,
And those roads in South Dakota that feel
around in the darkness. . .

MOYERS: You're a man of many paradoxes, Robert Bly. Light and darkness, love and death. Do you ever feel them—tearing at you?

BLY: Well, so what's wrong with a little insanity? (Laughs, playing dulcimer.) I suppose if you feel it, yeah. Eventually—I don't know; I don't know anything about it. But if I ever became wise, they'd all make a harmony, a terrific harmony. And the wider the grasp out to the edges, the bigger the harmony would be. I'll give you a poem by Rainer Maria Rilke, on the idea of trying to go out as far as you can to the edges, hm?

I live my life in growing orbits
Which move out over the things in the world.
Perhaps I can never achieve the last,
But that will be my attempt.

I am circling around God, around the ancient tower,
and I have been circling for a thousand years,
And I still don't know if I am a falcon, or a storm,
or a great song.

(Laughs.) That's Rilke when he was about seventeen years old. (Plays dulcimer.)

The body is like a November birch facing the full moon
And reaching into the cold heavens.
In these trees there is no ambition, no sodden body,
 no leaves,
Nothing but bare trunks climbing like cold fire!

My last walk in the trees has come. At dawn
I must return to the trapped fields,
To the obedient earth.
The trees shall be reaching all the winter.

It's a joy to walk in the bare woods.
The moonlight is not broken by the heavy leaves.
The leaves are down, and touching the soaked earth,
Giving off the odor that partridges love.

MOYERS: From Kabekona Lake in Minnesota and Cooper Union in New York, we've been listening to Robert Bly. I'm Bill Moyers.

BILL MOYERS' JOURNAL (air date: February 19, 1979)

TRANSMISSION

PART III: DISTRIBUTION

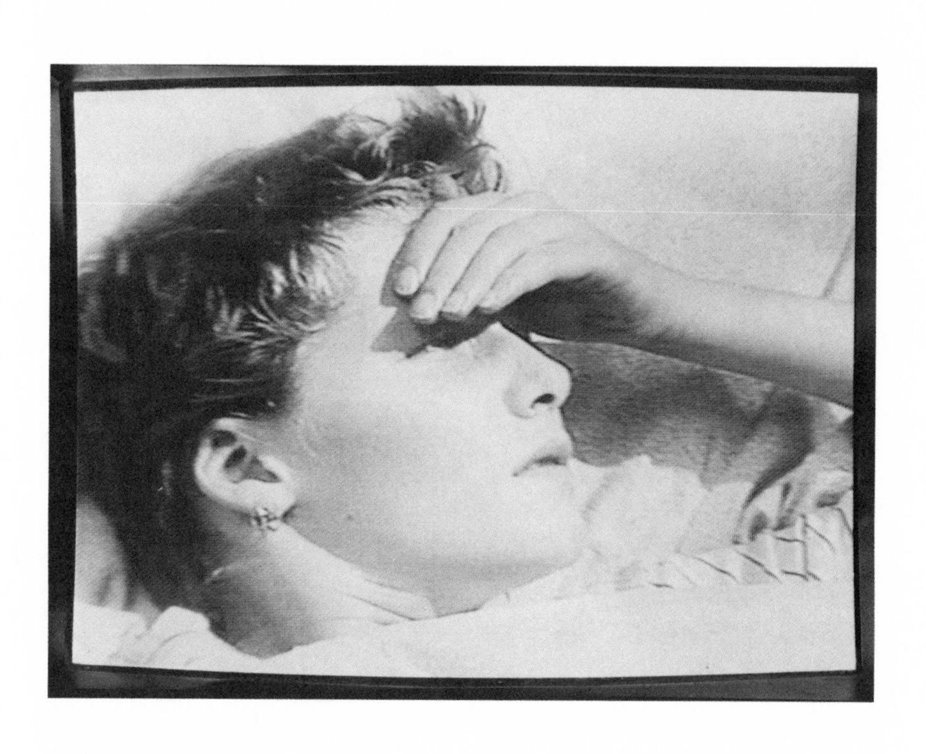

TUBE WITH A VIEW: BRITISH CHANNEL FOUR

Kathleen Hulser

"THE FOURTH CHANNEL is expected, by providing a favoured place for the untried, to foster the new and experimental in television."—Fourth Channel Programme Policy Statement

America has an abundance of television channels broadcasting a narrow range of themes and opinions. Britain offers only four channels but on its newest one, a role call of the excluded—from feminists to workers to Asians to Marxists to punks—address a national audience. Channel 4, which began broadcasting in November 1982, relies on outside independent producers to provide a programming mix that attempts to bring new voices to the screen in new ways. A sketch of the British context helps place the Channel 4 phenomenon in perspective.

Britain has two public television channels, BBC 1 and BBC 2, financed through a yearly tax on TV sets. Advertising finances ITV 1, the commercial channel. The BBC's staff produces its own programming but ITV 1's fare is supplied by outside production companies such as Granada or Thames Television. These independent contractors bid on franchises to make programs. Channel 4 is administered by the Independent Broadcasting Authority which also supervises ITV companies' ad revenues which mostly fund C4.

The important framework to understand when considering the evolution of Channel 4, is that broadcast television in Britain is deemed an issue of public policy. Various constituencies—from communications radicals such as the Glasgow Media Group, to labor activists, to Blacks and Asians, to feminists, to members of audiences never served—made proposals and gave testimony to government commissions, and wrote newspaper editorials and articles advocating political and aesthetic diversity as Channel 4's aim. Thus the various ideas incorporated in the final C4 charter were forged in a political arena.

The players

The key Labour Party panel, appointed in 1974—the Annan Committee—laid the groundwork for the Channel, and included representatives of many of the above-mentioned groups. For example, member Anthony Smith (later head of the British Film Institute), proposed a National Television Foundation dedicated to "publishing" neglected views and innovative form. His structural concerns—drawn from his awareness of how institutions shape the media, and from his production experience—were to ensure that an entrenched bureaucracy invested in the establishment didn't gain control of the new channel. The NFT, later dubbed the Open Broadcasting Authority, was nearly swamped in 1979 by the conservatives' plans for another commercial channel, but finally survived as a commitment to broadcasting a broader spectrum of political and social views. The other main elements which lasted through the political changes of the 1970s were: a structure of commissioned editors who obtained programming from large and small independents and the philosophy of showing work that didn't balance politically and generationally, thereby addressing particular, rather than large diffuse audiences.

Finally, the selection of Jeremy Isaacs as channel head was a mandate for a fresh approach. A producer and television administrator with experience in both commercial and public television, Isaacs had demonstrated his integrity when he quit ITV giant, Thames Television over the censorship of a documentary. His extensive background in current affairs included regular series and "The World at War".

The Programs

Since its debut in 1982, Channel 4 has amassed a 75-hour-a-week schedule assembled on a £100 million ($112 million) budget with 200 employees on staff. Program costs average £30,000 an hour ($33,600) which is low for Britain. Over a dozen commissioning editors contract series and specials, stressing co-productions with mostly European counterparts for big projects, and producing books and pamphlets to back up the 15 per cent of its material which must be educational (according to an IBA rule). A Welsh language service, Sianel Pedwar Cymru, fills the 4 slot in Wales, and is said to be lively and popular amongst the Welsh who waged a civil rights battle for years to obtain television in Welsh.

Despite the stirring talk of innovation a good deal of Channel 4's programming consists of television—a schedule operating in the British public service mode with more non-fiction, "infotainment," and high culture than one might see on PBS. But perhaps 10 to 20 percent of the channel is devoted to fare that is highly unusual—and risky by any standard of the television industry. It's these ventures that provide a model, or at least food for thought to those interested in vitalizing America's broadcast channels.

The leading weekly current affairs slot is filled by "Diverse Reports," a somewhat more conservative successor to the cancelled "Friday Alternative." Produced by David Graham, a man whose news sense is intelligently shaped by communications theory, "Friday Alternative" had adopted a Gramscian approach to dissolve the consensus views dominating a conventional news agenda. To dodge the shadow of mainstream institutional influences in "FA" Graham organized advisor groups of "ordinary people" around the country which debated issues without any intervening filter of a professional news commentator.

"Film on Four" began in a spirit of cooperation with the British Film Institute on such outstanding projects as Peter Greenaway's *The Draughtsman's Contract,* and has continued by backing such talents as Jerzy Skolomowski for *Moonlighting,* and Laura Mulvey and Peter Wollen for *The Bad Sister.* Although these films commissioned by David Rose would fall under the genre "art cinema," rather than "experimental" C4's purchasing power and audience *has* catalyzed filmmaking on British themes by newer directors.

Cutting edges

It's usually around 11 pm that Channel 4 enters a realm of experimental and politically fresh work that distinguishes it from broadcasting as we know it. Smaller independents who produce in workshops (media centers), not only have a forty-to-fifty week program slot called "The Eleventh Hour," they also have a privileged working relationship to the Channel. Instead of commissioning specific films and videos from them, C4 subsidizes the workshops themselves with from £35,000 ($39,200) to £150,000 ($168,000) per year flowing to selected workshops franchised under The Workshop Declaration. Many notable works have emerged from the workshops and have been aired on "The Eleventh Hour." "New Waves" offered such recent features as *Ghost Dance,* a film about cargo cults with Pascale Ogier and Jacques Derrida; and the painful political awakenings of Menelik Shabazz' *Burning an Illusion.* Viewers of "Africa on Africa" could sample Sembene Ousmane's films or Mauritanian Med Hondo's work. In "Ireland," a series of portraits recast the troubled history of Northern Ireland from an insider's perspective.

Women

The question of women's role at the Channel has ignited controversies from the day the Queen assented to the C4 charter. At the outset, Isaacs and senior editor for actuality, Liz Forgan (ex-women's editor at *The Guardian*), supported a twice monthly current affairs series that would examine the world through women's eyes. While the resulting program, "Broadside," broke new ground by attending to women's issues, it suffered from debilitating internal disputes. Finally, the group self-destructed. Analysts of "Broadside's" capsize point out that the historic exclusion of women from broadcasting weakened the production team. Staffing at C4 and producing companies remains so uneven that The Women's Film, Television and Video Network is partially funded by C4 to foster more women's participation at all levels of production and policy. And critics note that although Channel 4 has hired many print journalists, a production group recruited from the ranks of feminist monthly *Spare Rib* failed to obtain a piece of the current affairs action.

Another current affairs series in prime time "20/20 Vision" by Gambles and Milne (a women's producing team), has been moving towards more in-depth mini-documentaries, presenting, for example, a controversial three-part series on child sex abuse in fall 1984. "Stand Your Ground," a one-time prime-time series, dealt with self-defense and won praise for its treatment of violence against women.

Many women produce in the independent sector, and several women's series have occupied "The Eleventh Hour." "Pictures of Women," a six-part series produced in video, was a landmark allocation of funds to a feminist perspective. Covering the roots and ramifications of modern sex and gender, the series juxtaposed public and private attitudes through skits, documentary and graphic collages.

But some members of the women's community feel the Channel's approach smacks of tokenism. They are afraid that the hoopla opening of "Pictures" indicated that the Channel felt it had "done feminism." Caroline Spry of Four Corners, a film workshop partially funded by C4, says: "I don't feel that others [except for Alan Fountain, editor of 'The Eleventh Hour'] at Channel 4 are very aware or supportive of workshops, and they all play it pretty safe with women." She was disappointed that a series of 16 shorts Four Corners produced hadn't run during prime time, rather than as inserts between "Eleventh Hour" films. The "New Waves" series, an important commission for independents, consisted entirely of men's films. But when the C4 women's record so far is seen in an American context it is impressive in terms of topics and penetration of the program schedule.

Multi-cultural programming

What about the minorities cited in Channel planning documents? This area of programming is generally thought weak. Two magazine shows—

"Black on Black" and "Eastern Eye"—alternate weekly. One evening's "Black on Black" run in January 1985 managed to run ethnic perspectives into a slick and superficial show-biz grab-bag, mixing entertainment with commentary on the "white news." It has since been cancelled. In late 1984, a new commissioning editor for multicultural affairs was named. Newspaper articles greeted playwright Farrouk Dhondy's appointment as a promise of better things. Simon Blanchard, head of the Independent Film and Video Makers Association (IFVA), and editor of a book of alternative views on Channel 4, notes that even in the workshop sector, integration has been slow. He says a generation of experienced Black and Asian producers is just now reaching maturity (with the notable exception of Menelik Shabazz).

Conclusions

Channel 4 has opened a crack in the bland facade of television. This has occured through a combination of factors. First, it doesn't bear the sole responsibility for all public service broadcasting since BBC 1 and 2 contribute to the broad tasks of covering mainstream culture and providing education. Second, the structure of outside commissions means less danger of calcified staff orthodoxies. The preference shown for independents, large and small, plus the workshop funding, ensures that C4 is aware of and open to new ideas. And third, by American public broadcasting standards, a reasonably adequate and *steady* funding base makes it possible to build in an area.

Nevertheless, the bulk of Channel 4's money is spent on familiar television; the boundaries of the new are only gently pushed. In its overall treatment of the schedule C4 remains conventional, programming series in slots presumed favorable for known audiences. The schedule consists mainly of even length programs starting on or close to the hour, so viewers can conveniently switch in—or out.

Audience and press responses do count. Channel 4 watchers noticed that it made a half-turn in policy after its first year, and are concerned that early bold initiatives will whither as the Channel matures. But viewing figures are growing for "The Eleventh Hour," and for the Channel as a whole. While the 20-year-old BBC 2 draws a nine per cent audience share, Channel 4 has been creeping up to seven percent after two and a half years.

Simon Blanchard sums up the state of affairs saying: "There has been innovation: a range of work outside the political/social consensus has been shown—and gotten attention and debate. But Channel 4 *is* television." Ultimately, Channel Head Jeremy Isaacs has delivered what he promised in 1979: "a fourth channel that everyone will watch some of the time and no one will watch all of the time."

ASST. TO THE MGR.
AT THE BANK

THE TV LAB AT WNET/THIRTEEN

Marita Sturken

THE RELATIONSHIP OF ARTIST TO TELEVISION has always been complex. The central ideology of art is based on the primacy of the artist as an individual and the unique quality of the work; by its very nature, art questions conventions and familiar forms of language. Television, on the other hand, is dictated by a financial structure that requires a system of editorial hierarchies, serial formats, and the use of specific codes and conventions that are immediately comprehensible. Video artists have always had a love/hate relationship with television; it has the potential to be a powerful showcase for them, yet it represents the mediocrity of mass entertainment. Opportunities for artists in television have been few. While they have recently made forays into the arena of cable television, it was through public television that they were first able to infiltrate the airwaves. The network of public television differs from its commercial counterparts in programming for a specialized, upscale audience interested in culture, although it still requires large audiences to keep its funding intact. Idealistically, it should offer the kind of programming that fulfills television's potential as a powerful medium capable of bringing art and culture to a mass audience.

The failure of public television to meet the needs of a sophisticated audience and to adequately support and utilize the diverse independent film and videomakers of this country has been the subject of much debate over the past decade, practically all of which have stalemated on the fact that public TV is limited by its lack of sufficient funds. However, public television has not been without its inspired, experimental moments, and at a time when its very existence seems more threatened than ever before, those projects take on a new kind of historical aura. The TV Lab at WNET/Thirteen in New York, which officially closed its doors in 1984, was one of the most influential and highly visible programs for artists in public television. While its demise is indicative of changing political climates and funding priorities, its inception can be attributed to the communications ideology of the time.

The TV Lab was founded in 1972 with initial support from the Rockefeller Foundation and the New York State Council on the Arts (NYSCA). (In later years, it was also funded by the Corporation for Public Broadcasting and the Ford Foundation.) At this time, in the early 1970's, video art was characterized by a heady optimism about the possibilities of changing not only television, but the priorities of the communications industries; the belief that television could be democratic abounded. Spawned during the political unrest of the late 1960's, video was being used as a tool for social change and seen as an art medium that rejected the commoditizing of the art world. The rapid growth and optimism of the video community at that time is in large part attributable to the availability of unprecedented amounts of funding from NYSCA, the National Endowment for the Arts, and the Rockefeller Foundation (which gave initial funding to not only the TV Lab at WNET but also to the National Center for Experiments in Television at KQED in San Francisco, and to the Artists' Television Workshop at WGBH in Boston). In this context, the future of artists' public television looked bright and exciting.

The TV Lab was established as "a place where the broadest kind of artistic and technical exploration of the medium would be possible," according to Carol Brandenburg, who served as Co-Director of the Lab with David Loxton. "It was to be a television 'workshop' where artists and creative people from many disciplines could experiment with new forms and techniques." The Lab's mandate was to maintain an atmosphere where creative risk was possible, and to provide artists with access to broadcast equipment. In the realm of television, creative risk is a radical concept, and at a time when most artists had little or no access to editing equipment, the television studio at WNET represented a chance to use myriad sophisticated imaging devices to express aesthetic concerns.

During the TV Lab's first years, artists such as Shirley Clarke, Ed Emshwiller, and Nam June Paik produced tapes as resident artists. In 1974, the Artist-In-Residence Program was officially established, with five to eight artists chosen each year. The Lab also became involved in packaging some of this work for broadcast; in 1975 it produced "VTR (Video and Television Review)" a series hosted by curator Russell Connor. In 1979, this program became "Video/Film Review" to include independent film, broadcasting work regularly on Sunday nights.

Perhaps the most significant evidence of the TV Lab's influence is the comprehensive body of work that was generated under its auspices. Its first few years saw the production of several now-classic tapes, including Nam June Paik's *Global Groove* and Ed Emshwiller's *Scape-mates*. *Global Groove* is Paik's manifesto on universal communications, combining rock music, dancers, Japanese commercials, Charlotte Moorman playing Paik's *TV Cello*, and a chanting Allen Ginsberg in a rapid-fire collage that incorporated and subverted television techniques. *Scape-mates* is a surreal study of dancers in an electronic landscape. Its amorphous shapes, computer animation, and elusive fantasy quality made radical television, especially at the time.

The setup at the TV Lab first provided artists not only with an array of post-production equipment, including several artist-designed imaging machines such as the Paik/Abe synthesizer and the Rutt/Etra scan processor, but also with a studio space. For a number of years, the Lab was housed at Studio 46, where artists could shoot tapes. There, dancer Merce Cunningham with video artist Charles Atlas produced *Blue Studio*, a videotape in which Cunningham danced in the blue space of the studio. Atlas used a keying method to overlay various imagery onto the blue space, thus transforming Cunningham into many different environments, including one in which five Cunninghams dance at once. In 1977, however, the Lab refocused its efforts in response to the fact that most artists were interested in shooting their tapes out in the field; it closed Studio 46 and centered its efforts on its post-production facility.

While the Lab produced work by video artists and artists from other disciplines, it was also involved in supporting documentary work. Most independent documentaries make controversial TV in their exploration of subject matter than remains taboo on network television. TVTV (Top Value Television) was a collective of video pioneers who produced unusual documentaries in the Lab's early days. Their first production was *Lord of the Universe*, a portrait of teenage Guru Maharaj Ji that employed the loose intimate style of early street videotapes, and they went on to produce a notorious interview with fugitive Abbie Hoffman. Another Lab-supported project was *The Police Tapes*, by Alan and Susan Raymond, which was one of the first independent videotapes ever broadcast on commercial television. It is a harrowing, gritty documentary about a police precinct in the South Bronx, in which the Raymonds employed a verite style that gave a stark realism to the work.

The Raymonds went on to produce documentaries for commercial television, and they are not the only commercially successful documentarians who first produced works at TV Lab. Jon Alpert and Keiko Tsuno of Downtown Community Television (DCTV), a production center in New York's Chinatown, produced several important documentaries under the auspices of the TV Lab before going on to work for NBC. Their *Cuba: The People*, a document of a trip to Cuba in 1974, was the first documentary made with portable half-inch equipment to be broadcast on national television. Other DCTV productions, including *Health Care: Your Money or Your Life*, which dealt with the inadequacy of health care in America, explored subjects in a straightforward, candid style. Skip Blumberg, who was part of the innovative documentary collectives of video's early years produced several very popular documentaries at the Lab, including *Pick Up Your Feet: The Double Dutch Show,* about the inner city sport of double dutch jumping in New York City. Blumberg's intimate style and quirky subject matter combine to make energetic tapes that work well on television. In addition to supporting documentaries through the artist-in-residence program, the Lab also administered the Independent Documentary Fund beginning in 1977, producing independent documentaries many of which were broadcast in PBS's "Non-Fiction Television" series.

Just as the TV Lab helped to launch the careers of many of these documentary videomakers, participation in its program also served to provide a certain amount of credibility to video artists. While this career boost was first attributable to the fact that these video artists were gaining access to previously unavailable hardware, in later years—when other production facilities were also available—the status symbol of being a TV Lab artist-in-residence remained, proving perhaps that despite the ambivalence artists have toward TV, it still represents a powerful prestige in the video community. (It should be noted, however, that some of the TV Lab artists were a small, elite group. Many, such as Bill Viola, Nam June Paik, John Sanborn and Kit Fitzgerald, and Mitchell Kriegman, returned to do more than one production. Says Viola, "It's the place where I grew up in video. When you were accepted into the program, you made a quantum leap in terms of what was possible."

The works produced by these artists over the past decade form in many ways a sophisticated array of prototypes for a different kind of television—one that is intelligent, capable of reaching diverse audiences, and which explores new styles and controversial issues. Bill Viola's poetic and cerebral works, such as *The Reflecting Pool,* in which he explores life's cycles from birth to death; Edin Velez's impressionist essays such as *Meta Mayan II,* a portrait of the Guatamalan Indians which uses a slowed, elusive style to examine their gestures and character; Dan Reeves' *Smothering Dreams,* a personal exploration of social violence and the terror of combat which combines documentary footage, reenacted scenes, poetic commentary, and personal remembrances; and Robert Ashley's *The Lessons,* a densely layered, postmodern opera-for-television, are examples of the diverse styles such a production entity encouraged.

The structure of the video art community has changed drastically since the TV Lab first began its operations. A discouraged struggle for funding in what has become an increasingly expensive medium has replaced the optimism of those early years, and a proliferation of media arts centers exhibiting work and providing artists with production facilities has changed the role and importance of public television programs. Ironically, though, in the 1980s more and more video artists are perceiving their tapes as "works for television" rather than "video art." Programs which use commercial facilities off hours to produce artists' works have proved more cost-effective than the high overhead of public television setups, and some funding sources have moved on to other fields. While it is important to acknowledge that there remain basic schisms between the way artists work and television functions, it is possible to see the demise of the TV Lab and similar entities as that of a good idea whose time had come.

THE WGBH NEW TELEVISION WORKSHOP

Susan Dowling

SINCE ITS OFFICIAL INCEPTION in 1974, the WGBH New Television Workshop (Workshop) has continually sought to make the tools of television accessible to artists working in a variety of disciplines. It continues to operate from a firm and unchanging philosophical base: The creation of innovative, experimental programming produced in collaboration with independents. It takes continuing advantage of the medium's unexplored potential to present material of technical and creative excellence in entirely new ways.

In its first years the Workshop offered commissioned artists ready use of low cost half inch video equipment, free from the pressures of conventional production scheduling and broadcast deadlines. It existed to take chances, to explore risks, to fail as well as succeed in an environment free from the burdens and demands of high cost, high quality technical production. In the first year more than seventy-five artists from the fields of dance, drama, video art and music streamed through the Workshop doors to try new forms and concepts from and for the medium.

Under the direction of Dorothy Chiesa the Workshop's half inch "studio" began life in an abandoned movie house just outside Boston. Chiesa worked to raise funding from the Rockefeller Foundation, the National Endowment for the Arts, the Corporation for Public Broadcasting, and the Massachusetts Council on the Arts and Humanities. In 1976 over 300 independents strode in and out with fresh ideas for high-risk, low cost programming.

Early experiments and projects by Fred Barzyk were what inspired the creation of the Workshop. These efforts, under the Rockefeller supported Artists-in-Television Project, included works by Allan Kaprow, Nam June Paik, Otto Piene, James Seawright, Thomas Tadlock and Aldo Tambellini. They collaborated with television technicians in search of new ways to explore television as an electronic art form on "The Medium is the Medium," aired nationally in 1969. The program also served to focus attention on the independent artist as prime mover in television program innovation.

In 1967-68 there was a pre-workshop series called "What's Happening, Mr. Silver?." This one hour weekly program was hosted by David Silver, a professor of English at Tufts University. Each show was a collage of intellectual, visual, aural and emotional bombardment.

In 1969, Korean-born Nam June Paik, "the George Washington of Video," and his Japanese engineer Shuya Abe, came to Boston with five junk television sets, a set of giant magnets, miles of masking tape, rubber boots (to prevent shock) and hand-drawn schematics for the world's first videosynthesizer. With financial and creative support from WGBH, Paik brought his dream, the synthesizer, to life, and broadcast his four-hour "Video Commune - The Beatles: From Beginning to End." All images on the show—surreal landscapes (crushed tin foil), eerie abstractions (shaving cream), bursts of color (wrapping paper)—were transmogrified by the synthesizer at the very moment of broadcast: "live" television at its most unexpected. The videosynthesizer also gave impetus to Ron Hays, whose metier is the creation of video image 'scores' to classical music. Hays directed the Music Image Workshop at WGBH to encourage other video artists to work in the field.

WGBH continued to provide independents with access to broadcast for experiments through the '70s. A television was "The Very First On-The-Air Half Inch Videotape Festival Ever," in 1972. The Workshop invited anyone they could find who was working with half inch portapak equipment into studio A with their tapes, monitors and decks. The four hour festival included documentaries; political tapes by people who saw this new medium as a revolutionary tool; humorous and dramatic work by high school and college students as well as work by video artists.

With public awareness of video art growing, the Boston Symphony Orchestra collaborated with WGBH to rethink "the rather static visual nature of orchestral concerts." Together a group of artists were commisioned to make the BSO telecasts more interesting to the eye. The result, "Video Variations," was an hour special featuring the work of eight artists experimenting with television images suggested by orchestral music—a far cry from coventional concert coverage. (1973).

Another landmark achievement for the Workshop was "Video: The New Wave," aired nationally on PBS in 1974. Written and narrated by Brian O'Doherty, the program was simultaneously a retrospective and a preview of video artistry, assembling works produced at the Workshop and other pieces produced independently. Artists included were: Otto Piene, Doug Davis, Bill Etra, Stan VanDerBeek, Rudi Stern, David Atwood, Ron Hays, Stephen Beck, Dan Hallock, Nam June Paik, William Wegman, Frank Gillette, Peter Campus and Ed Emschwiller.

Also in 1974, the Dance Project of the Workshop was formed under Nancy Mason Hauser to explore ways of breaking down the barriers between the worlds of dance and video. This was the beginning of an on-going mandate to work with choreographers in developing video dance which has been carried on by former dancer Susan Dowling since 1979. Bernice Olenick joined the Workshop on a part-time basis in 1981. Her focus has been a new music series, "Soundings".

The Workshop gave up its studio in 1978, giving most of its equipment to BFVF (Boston Film and Video Foundation.) The current scene at the WGBH New Television Workshop is, for worse and for better, not as crazy or reckless as those early days and the financial strains are more difficult than ever. But the never-ending spirit and perserverence modeled after Barzyk continues. In addition, there is the renewed energy and interest on the part of artists to work in video. Independents from all disciplines (writers, visual artists, performance artists, musicians and choreographers) still receive commissions and technical support.

We want to help build a body of work in new video art fields and encourage other stations and media centers to do the same. We want to persevere in introducing these vanguard works to the mainstream. We want to compete with excellence in artistic vision, craft and content. The Workshop may be the only on-going operation of its kind in the world dedicated to working with artists in developing new video art forms. 1984 will celebrate our tenth year of commissioning artists to create new works for television.

Besides the on-going operation of the Workshop, we want to encourage as many television stations and media centers (world wide) to work with artists on projects of all sizes and budgets, broadcast and non-broadcast. More places equal more funding, more work and more exposure.

The Workshop has worked with a long list of artists. Among them are:

In Dance:
Rudy Perez, Gus Solomons, Jr., Dawn Kramer, Remy Charlip, Twyla Tharp, Dan Wagoner, Trisha Brown, Louis Falco, Meredith Monk, Marta Renzi, Douglas Dunn, Lisa Fox, Deborah Hay, Karole Armitage, Bill T. Jones, Arnie Zane
In Drama:
Writers - Kurt Vonnegut, Jr., Jean Shepard, Mary Feldhaus-Weber, Charles Johnson, Ned White, Denis O'Neil
Performers - Lily Tomlin, Dan Aykroyd, Glynn Turman, Gilda Radner, Matt Dillon, James Broderick, George Coe

In Visual Arts:
Charles Atlas, James Byrne, Patrick Ireland, John Erdman, Willem Dafoe, Nam June Paik, Otto Piene, Stan VanDerBeck, Ron Hays, Ros Barron, Betsy Connors, William Wegman, Fred Simon, Lee Krasner, Peter Camp-pus, Andy Mann, Bill Viola, Joan Logue, James Benning
In Music:
John Cage, Michael Colgrass, Ralph Shapey, Lucas Foss, John Driscoll, Joan Tower, Earle Brown, Rhys Chatham, Laurie Anderson, Jeffrey Lohn, Peter Gordon, Ashley, Monk, Springsteen

The *WGBH New Television Workshop* is the only operation of its kind in the nation which is dedicated to working with artists and creating new works for television. The Workshop has the passion, committment & expertise for continuing on a long term basis innovative arts programming in all media. (Mixed-media, inter-arts, and performance art projects are in pre-production)

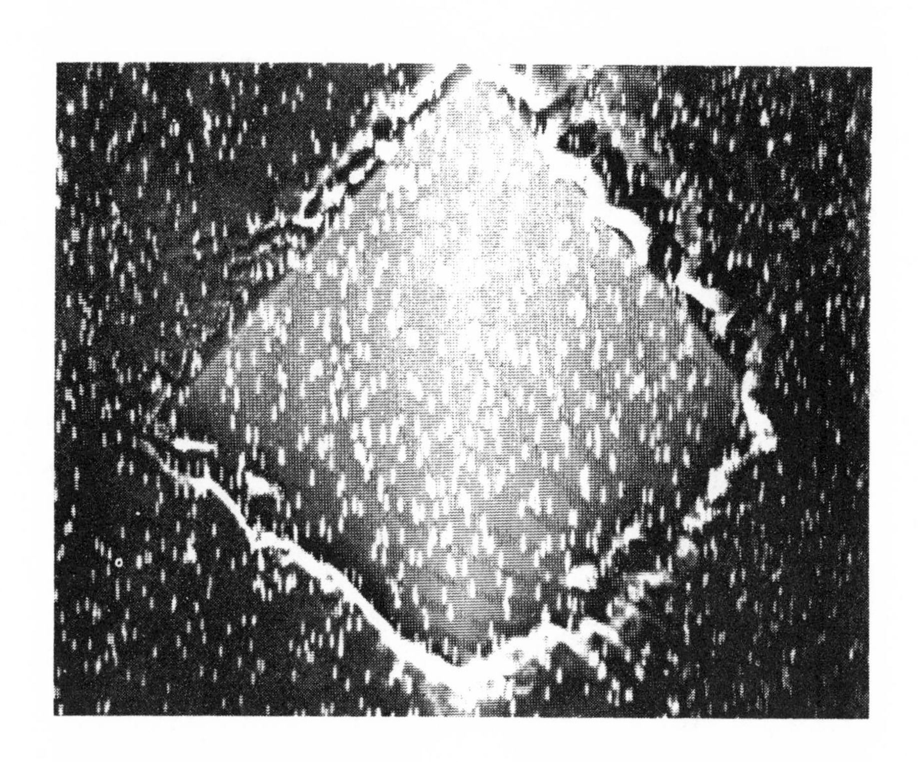

THE NATIONAL CENTER FOR EXPERIMENTS IN TELEVISION KQED

Joanne Kelly

THE CURRENT INTEREST in pioneering video artists and the media centers that support them is an indication of just how far video has come as an art form. There is now a history to refer back to—a lost history to rediscover.

The San Francisco Bay Area played a prominent role in developing television as an art form in the late 1960s and early 1970s. The video "scene" flourished here, especially through the work of groups like the National Center for Experiments in Television, Video Free America, Optic Nerve, Ant Farm, TVTV, and the De Saisset Gallery. This compilation is a profile of one of the most important early video centers, the National Center for Experiments in Television. Like the TV Lab at New York's WNET-TV and the New Television Workshop at Boston's WGBH-TV, the NCET was one of a handful of media centers affiliated with a public television station.

From 1967 to 1974 the National Center for Experiments in Television at KQED supported significant research and development for the field. Over its seven year lifespan it received $600,000 from the Rockefeller Foundation, substantial funds from the Corporation for Public Broadcasting, as well as the various benefits of an affiliation with a public television station in a major market. From this secure organizational base, NCET fulfilled its mandate of unrestricted video experimentation.

In its first year, under the direction of Brice Howard, NCET stressed a collaborative, interdisciplinary approach with a poet, a filmmaker, a novelist, a painter-sculptor, and a composer working with a television director for KQED. One of the most interesting programs from this period was Joanne Kyger and Bob Zagone's *Descartes*. Other collaborations with prominent artists, such as theater director Tom O'Horgan of "Hair" fame, occurred as well, yielding innovative programs like *Heimskringla*. At this time, KQED itself was a hotbed of radical programming.

In its second phase, the National Center moved out of KQED into its own loft space. The makers were not involved in collaborative projects as before. Instead, NCET evolved into a small group of individual videomakers, each with his own interdisciplinary background, each exploring new ways of thinking. It was at this time that Stephen Beck built his direct video synthesizer. Although several programs were aired on KQED during this period, including the Emmy award winning *Lostine* by Willard Rosenquist and Bill Roarty, there was never the pressure of producing for a broadcast date. Experimentation was viewed as an informative end in its own right; process triumphed over product.

Center artists revolutionized the environment that video was viewed in. In a time when video projectors were not in common use, Don Hallock's large kaleidoscopic *Videola*, essentially an environmental sculpture using conventional TV images and mirrors, presented new possibilities for viewing untraditional television. Center artists also went on to challenge the sense of time that television inherited from film—in particular, in the work of Bill Gwin, which, according to the artist, would ideally be presented "in a loop, running continuously. There would be no beginning, no middle, no end, and no particular duration. . . in much the same way a person spends time with a painting. . . ."

In the third and last phase of the Center, video art education at the University level was a prominent concern. The Center moved to Berkeley, establishing a video art workshop through University of California faculty member Willard Rosenquist. Other satellite centers at Rhode Island School of Design, Southern Methodist University, and Southern Illinois University at Edwardsville each contributed to this educational thrust.

In 1974, the Center as a place for aesthetic exploration began to dissolve and, by 1975, individual video artists were working on their own without the Center's organizational support. In an interview with Johanna Gill, Bill Gwin spoke fondly of the past: "It [the Center] was lucky for me because I learned how to use things in a very slow and unpressured way. There's no place like it anymore, which is a problem."

Indeed it is a problem. Today's non-profit media centers have jumped into the marketplace, hungry for commercial revenues from the use of their video artist's facilities. This current policy pits artists against industrial and commercial makers within the walls of alternative television centers. It is a far cry from the secure haven that the Center had provided for artists, as they built an award-winning body of synthesized/processed video art works. The Center is a reminder that original, innovative work needs freedom to grow and develop. "Process over product" as a philosophy certainly can be abused, but the visionary commitment the Center had for fostering unrestricted experimentation may be its most powerful legacy.

VIDEO: A BRIEF HISTORY AND SELECTED CHRONOLOGY

Barbara London

OVER THE LAST TWENTY YEARS, independent video has developed in response to the spirit of the time, at a pace determined by the availability of funding and equipment. For many years the individual artist could not afford broadcast quality equipment. Then in the mid-1960's, portable video hardware, similar in function but technically less sophisticated than professional equipment was developed for the home market, bringing video within reach of the individual. Video was no longer the exclusive domain of the television studio.

Television, which became a commercially viable industry after World War II, is the transmission of audio-visual images in the form of electronic signals to the home receiver. "Video" refers to both the equipment—the camera; the recording and editing decks; and the monitor, or television set—and the electronic signal. Television was restricted to live or prefilmed programs until the Ampex Corporation released the first videotape in 1957. The 1960's, when videotape was being adopted by the networks, was a time of social ferment marked in the United States by civil rights marches, the women's movement, and antiwar activities. Simultaneously traditional forms in the arts were being challenged by various expressions of the avant-garde, including kinetic art and multimedia Happenings, body art and nonnarrative film. In 1963 Nam June Paik and Wolf Vostell separately made iconoclastic sculptures by doctoring the internal circuitry of television sets. Both artists showed their television sculptures with distorted pictures on the screens in New York and West Germany, and both made films based on the misshapen electronic imagery.

When the Sony Corporation released the first inexpensive, nonbroadcast-quality, portable black-and-white camera in 1965, the individual gained access to the medium. At the beginning video was one of many tools that artists in New York, Los Angeles, Vancouver, London, Buenos Aires and Tokyo used in their mixed-media investigations. An information vehicle with recording and instant playback capabilities, video was used on both political and personal projects. Working with the rudimentary new video cameras, at first without editing equipment, artists were not bound by the conventions of television or traditional art. Some took the hardware outdoors to make "street tapes." Others made multi-monitor, sculptural installations or video performance works, using live cameras to create spatial and temporal projects in response to the particular environment. Given the scarcity of the equipment at the start, it was often easier to develop theories about video than to put the ideas into practice. The artists' philosophies were often related to Marshall McLuhan's broad approach to mass culture, or to John Cage's avant-garde notions about chance and time. Gradually the artists became familiar with the unique characteristics of the medium: its intimate scale, sense of time-based immediacy, electronic imagery, and light-emitting screen. Two early exhibitions, "TV as a Creative Medium" at the Howard Wise Gallery in New York in 1969 and "Vision and Television" at the Rose Art Museum outside of Boston in 1970, asserted that video was a viable art form.

In the late 1960's in North America, funding institutions began supporting video (the New York State Council on the Arts, the National Endowment for the Arts, the Canada Council, and the Rockefeller Foundation). The early grants went to facilities where small-format production and post-production equipment was made available to independents, and to screening programs such as the Kitchen's in New York. Grants were also awarded to major public television stations, whose producers—first Fred Barzyk at WGBH-Boston and Brice Howard at KQED-San Francisco, and later David Loxton at WNET-New York—initiated experimental video production and broadcast programs. Concurrently Gerry Schum was doing similar work with television stations in West Germany. Through these public television projects some fortunate artists gained access to broadcast quality video hardware and to the airwaves.

Initially there was little information on how to operate the new portable cameras and decks, which were being constantly modified by manufacturers. In the late 1960's such early groups as the Videofreex, Global Village, and Shirley Clarke's T.P. Video Space Troupe in New York; Video Free America in San Francisco; Intermedia in Vancouver; and Telewissen in West Germany held workshops as well as open screenings. But videomakers worked largely in isolation in their different cities. Then New York's Raindance group published *Radical Software*. The magazine disseminated both technical information and news about individual artists' productions, and helped to unify independent videomakers in North America. Simultaneously, the founders of Video Inn in Vancouver brought artists together through their publication of the *Video Exchange Directory* and their international Matrix conferences.

Once a modest amount of independent video existed, federal, state, and private support bolstered video exhibition and screening programs in museums, libraries, and alternative exhibition centers. The development of the 3/4-inch cassette in 1972 prompted the beginning of video distribution by Electronic Arts Intermix and Castelli-Sonnabend Tapes and Films in New York, Art Metropole in Toronto, Studiogalerie Mike Steiner in Berlin, and London Video Arts in London.

It is the personal point of view, made possible by the portable camera, that has distinguished artists' video from commercial material. Initially, network engineers would not accept on technical grounds anything made with the simple portable equipment; but the artists' spontaneous approach to the medium and their improvised subjects eventually influenced the style of television. Although many artists initially turned their backs on broadcasting, those with an activist or documentary approach considered television a desirable outlet. The counter-culture group Videofreex began regular programming on low-power television in Lanesville, New York, in 1971. Others such as Byron Black, Clive Robertson, and Ian Murray in Canada, and Jaime Davidovich and later Paper Tiger Television and the Artists' Collab in New York, began their own public-access cable series. Then in 1972 the Time Base corrector became available. This device, which could electronically correct deviational errors in the video signal caused by inconsistencies in equipment, meant that independent video could now be broadcast by national television. TVTV (Top Value Television)'s coverage of the 1972 political conventions and Downtown Community Television Center's report on Cuba's people were among the first independent documentary productions made with "nonprofessional" video equipment to be broadcast, both by WNET.

During the 1970's artists working with the new medium had great freedom. Unhampered by tradition and ignored by most mainstream art writers, videomakers did not have the demanding pressures (or potential for sales) that painters and sculptors had. Working with portable equipment, videomakers had absolute control of the writing, directing, and performing in their projects. Some, utilizing the intimate relationship between the viewer and the monitor, made personal narratives. Vito Acconci's work was boldly autobiographical, while William Wegman's humorous vignettes were focused on his Weimaraner dog, Man Ray. Canadian Colin Campbell acted out the different personae of his stories, while artists like Vera Frenkel in Canada and Tamara Krikorian in England exhibited their tapes with props related to their narrative tales.

Some artists enriched their performance with video. Joan Jonas composed metaphorical tableaus, using video to provide points of view normally not available to viewers during a performance, while German artist Ulrike Rosenbach created feminist rituals. At the same time, Charles Atlas and Merce Cunningham and Amy Greenfield developed a new genre of video dance, choreographing movement for the area framed by the video camera and the television screen.

Others have used video to create environmental studies with perception. In 1969 California artist Bruce Nauman carefully positioned a surveillance camera and monitor in a specially built narrow corridor so that viewers became disoriented when unable to confront their own images on the screen. Peter Campus developed psychological situations in which viewers could probe their elusive and exaggerated portraits projected onto the gallery wall. In his projects using a camera, monitors, and two-way mirrors, Dan Graham explored the process of viewing and being viewed; and conceptual installation projects were developed by Buky Schwartz in Israel, Richard Kriesche in Austria, and Catherine Ikam in France. Bill Viola has made poetic investigations of perception in both videotapes and video installations.

Beginning in the late 1960's, "cinematic" multimonitor installations were developed. Frank Gillette's projects are panoramic close-ups of luxuriant vegetation organized by abstruse theoretical constructs. Davidson Gigliotti has made serene, tripartite, pastoral works related to ninteenth-century landscape paintings, while Juan Downey first applied his fluid camera style to the examination of native Latin American culture. Beryl Korot languorously wove haunting images back and forth across the four monitors of her proscenium tableau in *Dachau*, as did Marie-Jo Lafontaine with her boxing scenes in *Round Around the Ring*.

From the beginning many artists explored video electronics, continuing the tradition of kinetic art and the computer films of Scott Bartlett and John and James Whitney. Among these are the former filmmaker Ed Emshwiller, Nam June Paik, Steina and Woody Vasulka, Dan Sandin and Tom DeFanti in the United States, and Toshio Matsumoto and Katsuhiro Yamaguchi in Japan. Some built their own image processors to achieve visual effects—such as the mixing together of live and taped images on a split screen—that had previously been possible only with expensive broadcast equipment.

The conventions of television itself—its genres, editing, and camera techniques—have been the subject of analytical works. For their early projects, video artists Telethon (William Adler and John Margolies) and Dara Birnbaum used excerpts from commercial television, editing them to correspond to the medium's fast commercial pace. On the other hand, Stuart Marshall in London and Tom Sherman in Toronto developed television's narrative form into their own analytical vocabularies.

Artists' video continues to be actively shown in a variety of settings: in the contemporary art biennial and the film/video festival, on cable and public television (on a limited basis), as well as rock clubs, alternative spaces, libraries, and museums. Today ongoing video exhibition and collection programs, as well as research centers for the study of the art form, exist at the Stedelijk Museum, Amsterdam; the Centre Georges Pompidou, Paris; The Museum of Modern Art and Whitney Museum of American Art, New York; the Long Beach Museum, Long Beach, California; the Institute of Contemporary Art, Boston; the Institute of Contemporary Art, London; the Vancouver Art Gallery, Vancouver; and the National Gallery of Canada, Ottawa. Most of the institutions that began video programs in the 1970s have continued their commitment to the medium.

Today as the critical vocabulary for video is being established, the strongest works in single videotape and installation formats are recognized as having an integral cohesiveness and integrity. At this point there are mature artists who have vast experience and who understand the potentials of the video medium. Because of their high technical standards, today it can cost more and take longer to complete a new project; unfortunately, this coincides with a time of cutbacks by federal and private art sources. However, the nexus between video and computer technologies on the level of home equipment is beginning to provide powerful new options. Independent videomakers occupy a special pioneering position now that the medium is about to emerge with the commercial film industry. These artists need to be encouraged for their visionary innovations.

The chronology that follows highlights some of the major events that have helped to shape independent video in the United States. Although institutions have provided the context for video, it is the artists' contributions that are of the greatest importance.

SELECTED CHRONOLOGY

1963
Exhibitions/Events

New York. TELEVISION DÉ-COLL/AGE by Wolf Vostell, Smolin Gallery. First U.S. environmental installation using a television set.

1964
Television/Productions

Boston. JAZZ IMAGES. WGBH-TV. Producer, Fred Barzyk. Five short visualizations of music for broadcast; one of the first attempts at experimental television.

1965
Exhibitions/Events

New York. ELECTRONIC ART by Nam June Paik, Galeria Bonino. Artist's first gallery exhibition in U.S.
NEW CINEMA FESTIVAL I (Expanded Cinema Festival), The Film-Makers Cinematheque. Organized by John Brockman. Festival explores uses of mixed-media projection, including video, sound, and light experiments.

1966
Exhibitions/Events

New York. 9 EVENINGS: THEATER AND ENGINEERING, 69th Regiment Armory. Organized by Billy Klüver. Mixed-media performance events with collaboration between ten artists and forty engineers. Video projection used in works of Alex Hay, Robert Rauschenberg, David Tudor, Robert Whitman.
SELMA LAST YEAR by Ken Dewey, New York Film Festival at Lincoln Center, Philharmonic Hall Lobby. Multichannel video installation with photographs by Bruce Davidson, music by Terry Riley.

1967
Exhibitions/Events

Minneapolis. LIGHT/MOTION/SPACE. Walker Art Center in collaboration with Howard Wise Gallery, New York. Travels to Milwaukee Art Center. Includes video works by Nam June Paik, Aldo Tambellini and others.
New York. FESTIVAL OF LIGHTS, Howard Wise Gallery. Exhibition of kinetic light works that include video works by Serge Boutourline, Nam June Paik, Aldo Tambellini, and others.
Rockefeller Foundation awards first video fellowship.
ELECTRONIC BLUES by Nam June Paik in "Lights in Orbit," Howard Wise Gallery. Viewer-participation video installation.

Television/Productions

Boston. WGBH-TV inaugurates artist-in-residence program with grant from the Rockefeller Foundation.
WHAT'S HAPPENING MR. SILVER? WGBH-TV. Host, David Silver. Experimental collage/information series in which several dozen inputs are mixed live and at random.
San Francisco. EXPERIMENTAL TELEVISION WORKSHOP, KQED-TV. Direc tors, Brice Howard and Paul Kaufman. Established with Rockefeller Foundation grant. In 1969 renamed National Center for Experiments in Television (NCET), funded by the Corporation for Public Broadcasting and the National Endowment for the Arts. Ends 1976.

1968

Exhibitions/Events

New York. BLACK: VIDEO by Aldo Tambellini in "Some More Beginnings," Brooklyn Museum. Organized by Experiments in Art and Technology.
ELECTRONIC ART II by Nam June Paik, Galeria Bonino.
INTERMEDIA '68. Theater Workshop for Students and the Brooklyn Academy of Music. Organized by John Brockman. Funded through the New York State Council for the Arts. Exhibition includes environmental video performances, light and film projections, videotapes. Video by Ken Dewey with Jerry Walter, Les Levine with George Fan, Aldo Tambellini.
IRIS by Les Levin. First shown publicly in artist's studio. Sculpture with six monitors and three video cameras, commissioned by Mr. and Mrs. Robert Kardon. Collection, Philadelphia Museum of Art.
THE MACHINE AS SEEN AT THE END OF THE MECHANICAL AGE, The Museum of Modern Art. Director of exhibition, Pontus Hultén. Exhibition includes video art, particularly Nam June Paik's *Nixon Tape, McLuhan Caged,* and *Lindsay Tape* on unique tape-loop device.
TIME SITUATION by David Lamelas in "Beyond Geometry," Center for Inter-American Relations. An installation using television monitors in exhibition sponsored by the Instituto Torcuato di Tella, Buenos Aires.
Washington, D.C. CYBERNETIC SERENDIPITY: THE COMPUTER AND THE ARTS, The Corcoran Gallery. Travels to Palace of Art and Science, San Francisco. Director of exhibition, Jasia Reichardt. Exhibition originated at Institute of Contemporary Art, London; American showing augmented by work selected by James Harithas. Includes video work by Nam June Paik.

Organizations

New York. BLACK GATE THEATER, for electromedia events, and *Gate Theater,* for experimental independent cinema. Founded by Aldo Tambellini.
COMMENDATION. Video production group. Original members: David Cort, Frank Gillette, Howard Gudstadt, Ken Marsh, Harvey Simon. Ends 1969.
YOUNG FILMMAKERS/VIDEO ARTS. Educational organization with training services, workshops, production facilities. Director, Roger Larson.
San Francisco. ANT FARM. Artists' media/architecture group. Founded by Chip Lord and Doug Michels; joined by Curtis Schreier in 1971. Other members include Kelly Gloger, Joe Hall, Hudson Marquez, Allen Rucker, Michael Wright. Disbands 1978.
LAND TRUTH CIRCUS. Experimental video collective. Founded by Doug Hall, Diane Hall, Jody Proctor. In 1972 renamed Truthco; in 1975, T. R. Uthco. Ends 1978.
Santa Clara, Calif. THE ELECTRIC EYE. Video collective. Founded by Tim Barger, Jim Mandis, Jim Murphy, Michelle Newman, Skip Sweeney. Ends 1970.

Television/Productions

New York. THE UNDERGROUND SUNDAE by Andy Warhol. Warhol com-

missioned to make a sixty-second commercial for Schraff's Restaurant.
San Francisco. SORCERY by Loren Sears and Robert Zagone. Experimental Television Workshop, KQED-TV. Live-broadcast program using special-effects imagery.

1969

Exhibitions/Events

New York. TV AS A CREATIVE MEDIUM, Howard Wise Gallery. First American exhibition devoted entirely to video art. Works by Serge Boutourline, Frank Gillette and Ira Schneider, Nam June Paik (With Charlotte Moorman), Earl Reiback, Paul Ryan, John Seery, Eric Siegel, Thomas Tadlock, Aldo Tambellini, Joe Weintraub.
Los Angeles. CORRIDOR by Bruce Nauman, Nicholas Wilder Gallery. Installation with video.

Organizations

Cambridge. CENTER FOR ADVANCED VISUAL STUDIES, Massachusetts Institute of Technology (MIT). Established for artists to explore art and technology. Founded by Gyorgy Kepes. Director, Otto Piene.
New York. CHANNEL ONE. Video theater offering comic programming featuring Chevy Chase. Director, Ken Shapiro. Technical Director, Eric Siegel.
GLOBAL VILLAGE. Begins as video collective with information and screening center. Becomes media center devoted to independent video production with emphasis on video documentary. Founded by John Reilly, Ira Schneider, Rudi Stern. Directors, John Reilly and Julie Gustafson.
RAINDANCE CORPORATION. Collective formed for experimental production. In 1971 becomes Raindance Foundation, devoted to research and development of video as a creative and communications medium, with screening program. Members: Frank Gillette, Michael Shamberg, Steve Salonis, Marco Vassi, Louis Jaffe; soon after, Ira Schneider and Paul Ryan, and then Beryl Korot.
VIDEOFREEX. Experimental video group. Members: Skip Blumberg, Nancy Cain, David Cort, Bart Friedman, Davidson Gigliotti, Chuck Kennedy, Curtis Ratcliff, Parry Teasdale, Carol Vontobel, Tunie Wall, Ann Woodward.

Television/Productions

Boston. THE MEDIUM IS THE MEDIUM, WGBH-TV. Produced by Fred Barzyk, Anne Gresser, Pat Marx. First presentation of works by independent video artists aired on television. Thirty-minute program with works by Allan Kaprow, Nam June Paik, Otto Piene, James Seawright, Thomas Tadlock, Aldo Tambellini.
New York. SUBJECT TO CHANGE, SQN Productions for CBS. Produced by Don West. Program of videotapes initiated by Don West with CBS and produced by Videofreex and other members of the video community. Videotapes produced on all aspects of the counterculture (alternate schools, communes, radicals, Black Panthers, riots, demonstrations, etc.). Never broadcast.

1970

Exhibitions/Events

New York. A.I.R. by Les Levine in "Software," the Jewish Museum. Curator, Jack Burnham. Eighteen-monitor video installation.
INFORMATION. The Museum of Modern Art. Curator, Kynaston McShine. Exhibition includes videotapes and installations from U.S., Europe, Latin America.
WAREHOUSE SHOW. Leo Castelli Gallery. Includes video installation by Keith Sonnier.
Plainfield, Vt. THE FIRST GATHERING: ALTERNATE MEDIA PROJECT, Goddard College. Media conference.
San Francisco. BODY WORKS. Museum of Conceptural Art. Videotapes by Vito Acconci, Terry Fox, Bruce Nauman, Dennis Oppenheim, Keith Sonnier, William Wegman. Organized by Willoughby Sharp. First video exhibition on the West Coast.
PHIO T. FARNSWORTH VIDEO OBELISK by Skip Sweeney, Intersection Theater, Multichannel video installation.
Waltham, Mass. VISION AND TELEVISION, Rose Art Museum, Brandeis University. Organized by Russell Connor. Works by Frank Gillette, Ted Kraynik, Les Levine, Eugene Mattingly, Nam June Paik (with Charlotte Moorman), John Reilly and Rudi Stern, Paul Ryan, Ira Schneider, Eric Siegel, Aldo Tambellini, Jud Yalkut, USCO/Intermedia, Videofreex, Joe Weintraub.

Organizations

Binghamton, N.Y. EXPERIMENTAL TELEVISION CENTER. Originally Community Center for Television Production. Production/post-production center emphasizing synthesized and computer-generated imagery. Directors, Ralph Hocking and Sherry Miller. In 1979 moves to Owego, N.Y.
Menlo Park, Calf. MEDIA ACCESS CENTER, Portola Institute. Alternative television resource emphasizing community and high school video programs. Original members: Pat Crowley, Richard Kletter, Allen Rucker, Shelley Surpin. Ends 1972.
New York. Creative Artists Public Service (CAPS) awards fellowships in video.
ELECTRONIC ARTS INTERMIX. Founded by Howard Wise after he closes his gallery; incorporated 1971. Explores video as a medium of personal expression and communication. In 1972 establishes editing/post-production facility. In 1973 begins Artists Videotape Distribution Service.
New York State Council on the Arts forms TV/Media Program. directors include Peter Bradley, Paul Ryan, Russell Connor, Gilbert Konishi, Lydia Silman, Nancy Legge, John Giancola.
PEOPLE'S VIDEO THEATER. Alternative video journalism collective emphasizing community video and political issues. Conducts weekend screenings in which the audience discussions are taped and replayed. Founded by Elliot Glass, Ken Marsh. Members include Judy Fiedler, Howard Gudstadt, Molly Hughes, Ben Levine, Richard Malone, Elaine Milosh, Richard Nusser.
San Francisco. MUSEUM OF CONCEPTUAL ART (MOCA). Alternative museum created for performance and multimedia art. Founded by Tom Marioni.
VIDEO FREE AMERICA. Video production group with post-production and

screening programs. Founded by Arthur Ginsberg, Skip Sweeney. Directors: Joanne Kelly, Skip Sweeney.
Syracuse, N.Y. SYNAPSE VIDEO CENTER (formerly University Community Union Video). Video production and post-production center. Directors include Lance Wisniewski, Henry Baker. Closes 1980.

Television/Productions

Boston. Nam June Paik and Shuya Abe develop Paik/Abe synthesizer while artists-in-residence at WGBH-TV.
VIOLENCE SONATA by Stan VanDerBeek, WGBH-TV. Live broadcast perfor mance with video-tape, film, and participation of studio and phone-in audience on theme of violence.
New York. Eric Siegel builds Electronic Video Synthesizer with financial assistance from Howard Wise.
San Francisco. Stephen Beck builds Direct Video Synthesizer 1, funded in part by the National Endowment for the Arts.

Publications

FILM AND VIDEO MAKERS TRAVEL SHEET (Pittsburgh: Museum of Art, Carnegie Institute). Monthly listing of artists' appearances, new works, events.
RADICAL SOFTWARE (New York: Raindance Foundation). Alternative video magazine and information channel for distribution and exchange of video works. Published 1970-74, vols. 1-2. Coeditors, Phyllis Gershuny and Beryl Korot. Publishers, Ira Schneider and Michael Shamberg.
EXPANDED CINEMA by Gene Youngblood (New York: E. P. Dutton). First publication to cover video art.

1971

Exhibitions/Events

Berkeley, Calif. TAPES FROM ALL TRIBES, Pacific Film Archive, University of California. Organized by Video Free America. Exhibition of videotapes by over 100 American artists.
THE TELEVISION ENVIRONMENT, University Art Museum. Produced by William Adler and John Margolies for Telethon. Circulates through American Federation of Arts.
New York. EIGHTH NEW YORK AVANT-GARDE FESTIVAL, 69th Regiment Armory. Director, Charlotte Moorman. Individual video projects by Shirley Clarke, Douglas Davis, Ken Dominick, Ralph Hocking, Nam June Paik, Eric Siegel, Steina and Woody Vasulka, Videofreex.
ELECTRONIC ART III by Nam June Paik and Shuya Abe with Charlotte Moorman, Galeria Bonino. Exhibition with Paik-Abe synthesizer.
Installation works by Vito Acconci, Bill Beckley, Terry Fox, William Wegman at 93 Grand Street. Organized by Willoughby Sharp.
PROJECTS: KEITH SONNIER. The Museum of Modern Art. Environmental video installation. Beginning of "Projects" exhibition program.
A SPECIAL VIDEOTAPE SHOW. Whitney Museum of American Art. New Amer-

ican Filmmakers Series. Organized by David Bienstock. Videotapes by Isaac Abrams, Shridhar Bapat, Stephen Beck, John Randolph Carter, Douglas Davis, Dimitri Devyatkiin, Ed Emshwiller, Richard Felciano, Carol Herzer, Joanne Kyger, Richard Lowenberg, Alwin Nikolais, Nam June Paik (with Charlotte Moorman), Charles Phillips, Terry Riley, Eric Siegel, Skip Sweeney, Aldo Tambellini, Steina and Woody Vasulka, WGBH-TV, Robert Zagone.

TEN VIDEO PERFORMANCES. Finch College Museum of Contemporary Art. Organized by Elayne Varian. Works by Vito Acconci, Peter Campus, Douglas Davis, Dan Graham, Alex Hay, Bruce Nauman, Claes Oldenburg, Nam June Paik, Robert Rauschenberg, Steve Reich, Eric Siegel, Simone Whitman.

PERCEPTION. Group of artists interested in alternative uses of video, explore video programming in conjunction with Electronic Intermix. Founded by Eric Siegel and Steina and Woody Vasulka. Subsequent members: Juan Downey, Frank Gillette, Beryl Korot, Andy Mann, Ira Schneider. Disbands 1973.

T. P. VIDEO SPACE TROUPE. Experimental workshop exploring two-way video. Founded by Shirley Clarke. Original members include Wendy Clarke, Bruce Ferguson, Andy Gurian. Disbands 1977.

WOMEN'S INTERART CENTER. Organization to create interdisciplinary collaboration involving writers, visual artists, performance artists, video artists. In 1972 begins post-production center. Offers workshops, produces videotapes, sponsors artists-in-residence. Director, Margot Lewitin. Video directors include Carolyn Kresky, Jenny Goldberg, Susan Milano, Ann Volkes, Wendy Clarke, Veronica Geist.

MEDIA EQUIPMENT RESOURCE CENTER (MERC), initiated by Young Filmmakers/Video Arts. Equipment loan service for artists and organizations. In 1977 reorganizes as access service with TV studio, equipment loan, and post-production divisions.

New Orleans. NEW ORLEANS VIDEO ACCESS CENTER (NOVAC). Founded through VISTA to provide video access to low-income community. Becomes production center with access.

Syracuse, N.Y. Everson Museum establishes first video department in a major museum, under direction of James Harithas. Video curators include David Ross, Richard Simmons. Department closes 1981.

Washington, D.C. National Endowment for the Arts initiates Public Media Program. Directors include Chloe Aaron, Brian O'Doherty. In 1977 becomes Media Arts Program.

Organizations

Chicago. VIDEOPOLIS. Video/resource teaching center. Founded by Anda Korsts. Closes 1978.

Ithaca, N.Y. ITHACA VIDEO PROJECTS. Organization for promotion of electronic communication. Director, Phillip Mallory Jones.

Lanesville, N.Y. MEDIA BUS. Founded by the Videofreex. Media center begins producing "Lanesville TV," weekly program about the community that is the first low-power television (LPTV) station. In 1979 Media Bus moves to Woodstock and operaties a post-production facility, distribution and consulting services, and produces programming for cable. Current members: Nancy Cain, Tobe Carey, Bart Friedman.

New York. ALTERNATE MEDIA CENTER. School of the Arts, New York University. Funded by the John and Mary Markle Foundation to explore the uses of broadcast telecommunications. Founded by Red Burns and George Stoney. Director, Red Burns.
THE ELECTRONIC KITCHEN. Screening and performance center for the electronic arts at Mercer Arts Center. Founded by Steina and Woody Vasulka, Andres Mannik. Subsequently The Kitchen Center for Video, Music and Dance. Video Directors include Shridhar Bapat, Dimitri Devyatkin, Carlota Schoolman, RoseLee Goldberg, Jackie Kain, Greg Miller, Tom Bowes, Amy Taubin.
OPEN CHANNEL. Organization for development of public access. Produces community programming, conducts workshops, school programs, and organizes talent pool of film and television professionals to produce public-access programming. Founded by Thea Sklover. Director of Programming, Lee Ferguson. Ends 1976.

Television/Productions

Boston. VIDEO VARIATIONS. WGBH-TV. Collaboration between Boston Symphony Orchestra and artists Jackie Cassen, Russell Connor, Douglas Davis, Constantine Manos, Nam June Paik, James Seawright, Stan VanDerBeek, Tsai Wen-Ying. Produced by Fred Barzyk.
New York. ARTISTS' TELEVISION WORKSHOP, WNET-TV. Established through efforts of Jackie Cassen, Russell Connor, Nam June Paik, with initial grant from New York State Council on the Arts to support experimental projects by independents.
New York City mandates public access as part of its cable franchise.
Providence, R.I. Satellite program of the National Center for Experiments in Television (NCET) established by Brice Howard at Rhode Island School of Design; also at Southern Methodist University, Dallas, and Southern Illinois University, Edwardsville.
Washington, D.C. ELECTRONIC HOKKADIM 1 by Douglas Davis, Corcoran Gallery of Art, and WTOP-TV. Live broadcast piece with two-way communication via telephone.

Publications

GUERRILLA TELEVISION by Michael Shamberg and Raindance Corporation(New York: Holt, Rhinehart and Winston). Manual of alternative television with graphic by Ant Farm.

1972

Exhibitions/Events

Minneapolis. FIRST ANNUAL NATIONAL VIDEO FESTIVAL. Minneapolis College of Art and Design and Walker Art Center. Organized by Tom Drysdale. Consists of workshops, screenings, panel discussion. Participants include Peter Campus. Russell Connor, Ed Emshwiller, Nam June Paik, Barbara Rose, Ira Schneider, George Stoney, Aldo Tambellini, Gene Youngblood.
New York. Peter Campus, Bykert Gallery. One-man show with video installations.

FIRST WOMEN'S VIDEO FESTIVAL. The Kitchen at Mercer Arts Center. Organized by Susan Milano. Includes work by Jackie Cassen, Maxi Cohen, Yoko Maruyama, Susan Milano, Queer Blue Light Video, Keiko Tsuno, Steina and Woody Vasulka, Women's Video Collective; and dance/video performance by Judith Scott, Elsa Tambellini.
NINTH ANNUAL NEW YORK AVANT-GARDE FESTIVAL, Alexander Hamilton Hudson Riverboat. Director, Charlotte Moorman. Includes special video projects by over fifteen artists.
Santa Clara, Calf. FIRST ST. JUDE INVITATIONAL OF VIDEO ART. de Saisset Gallery and Art Museum, University of Santa Clara. Organized by David Ross. Works by John Baldessari, Lynda Benglis, George Bolling, Douglas Davis, Taka Iimura, Videofreex, William Wegman.
Syracuse, N.Y. DOUGLAS DAVIS: AN EXHIBITION INSIDE AND OUTSIDE THE MUSEUM, Everson Museum of Art, with WCNY-TV. An exhibition with live telecast, "Talk Out!"
NAM JUNE PAIK. Everson Museum of Art. Tapes, installations, and performance, with Charlotte Moorman.

Organizations

Buffalo, N.Y. MEDIA STUDY/BUFFALO. Center for videotape production and exhibition. President, Gerald O'Grady; Video/Electronic Arts Curator, John Minkowsky.
New York. CASTELLI-SONNABEND VIDEOTAPES AND FILMS. Videotape distribution service. Founded by Leo Castelli and Ileana Sonnabend. Directors include Joyce Nereaux, Patricia Brundage.
DOWNTOWN COMMUNITY TELEVISION CENTER (DCTV). Educational and production organization. Founded by Jon Alpert, Keiko Tsuno.
FIFI CORDAY PRODUCTIONS. Organization to assist artists' production. Founded by Carlota Schoolman.
SURVIVAL ARTS MEDIA. Video collective emphasizing community education and health programs, programs on artists and artistic processes, and multimedia shows. Members include Gail Edwards, Howard Gudstadt, Molly Hughes, Ben Levine, Danny Luciano, Richard Malone.
Rochester, N.Y. PORTABLE CHANNEL. Video resource center with workshops, visiting artists series, equipment access, productions. Directors include Bonnie Klein, Sanford Rockowitz, John Camelio, Robert Shea, Tim Kelly.
St. Louis.DOUBLE HELIX Media Center with production and post-production facilities, audio/video workshops.
San Francisco. OPTIC NERVE. Documentary production collective producing political and social documentaries. Original members include Lynn Adler, Jules Backus, Jim Mayer, Sherrie Rabinowitz, John Rogers, Mya Shone. Disbands 1979.
TOP VALUE TELEVISION (TVTV). Independent documentary production group forms to provide alternative coverage of the Democratic and Republican conventions in Miami; the first use of half-inch videotape on broadcast television. Original production by Hudson Marquez, Allen Rucker, Michael Shamberg, Tom Weinberg, Megan Williams, and members of Ant Farm, Raindance, and Videof-

reex collectives. Other members of TVTV include Wendy Apple, Michael Couzens, Paul Goldsmith, Betsy Guignon, Stanton Kaye, Anda Korsts, Andy Mann, Elon Soltes. Disbands 1977.
Woodstock, N.Y. WOODSTOCK COMMUNITY VIDEO. Production center and resource for community video. Initiates local cable programming. Begins Artists' TV Lab, which moves to Rhinebeck in 1976. From 1975 to 1977 presents Woodstock Video Expovision, a festival of New York State artists. Founded by Ken Marsh. Members include Barbara Buckner, Bob Dacy, Gary Hill, Steven Kolpan, Elaine Milosh. Ends 1978.

Television/Productions

Boston. MUSIC IMAGE WORKSHOP. WGBH-TV. Project by Ron Hays using Paik-Abe synthesizer to produce tapes relating to music and video imagery.
THE VERY FIRST ON-THE-AIR HALF-INCH VIDEOTAPE FESTIVAL EVER: PEOPLE TELEVISION. WGBH-TV. Produced by Henry Becton with Fred Barzyk, Dorothy Chiesa. Live studio event including home viewer call-ins, tape screenings, and interviews with artists, engineers, business people, educators, students.
Chicago. Dan Sandin builds Image Processor, and eventually, with Phil Morton, makes plans available to artists.
New York. SCAPE-MATES by Ed Emshwiller, the Television Laboratory at WNET/Thirteen. Videotape with complex mixing of live actors and computer graphics.
THE TELEVISION LABORATORY AT WNET/THIRTEEN. Directors include David Loxton, Carol Brandenburg. Founded with grants from the Rockefeller Foundation and New York State Council on the Arts. First year initiates artist-in-residence program with Shirley Clarke, Douglas Davis, Ed Emshwiller, Nam June Paik.
San Francisco. ELECTRONIC NOTEBOOKS by Stephen Beck, KQED-TV. Series of tapes produced with Bill Gwin, Don Hallock, Warner Jepson, Bill Roarty, Willard Rosenquist.
Washington, D.C. The Federal Communications Commission (FCC) requires that all cable franchises have at least one public access channel.

Publications

BETWEEN PARADIGMS: THE MOOD AND ITS PURPOSE by Frank Gillette (New York: Gordon and Breach).
PRINT (New York: RC Publications). Special video issue. Guest editor, Robert de Havilland. Contributors: Fred Barzyk, Rudi Bass, Rose DeNeue, Bernard Owett, Sheldon Satin, Michael Shamberg.

1973

Exhibitions/Events

Los Angeles. WILLIAM WEGMAN. Los Angeles County Museum of Art. Exhibition of drawings and tapes.
New York. INTERNATIONAL COMPUTER ARTS FESTIVAL. The Kitchen at Mercer Arts Center. Organized by Dimitri Devyatkin. Includes music, poetry, film, video.

THE IRISH TAPES by John Reilly and Stefan Moore. The Kitchen at Mercer Arts Center. Installation with three channels and twelve monitors.
1973 BIENNIAL EXHIBITION. Whitney Museum of American Art. First inclusion of video in Biennial exhibition. Includes videotapes by seven artists and installation by Peter Campus.
TENTH NEW YORK AVANT-GARDE FESTIVAL. Grand Central Station. Director, Charlotte Moorman. Includes special video projects by over seventeen artists.
Syracuse, N.Y. CIRCUIT: A VIDEO INVITATIONAL. Everson Museum of Art. Curated by David Ross. Traveling exhibition of videotapes by over sixty-five artists. Travels to Henry Gallery, University of Washington, Seattle; Cranbrook Academy of Art Museum, Bloomfield Hills, Mich.; Kölnischer Kunstverein, Co logne, West Germany; Greenville County Museum of Art, Greenville, S.C.; and in 1974, Museum of Fine Arts, Boston.
FRANK GILLETTE: VIDEO PROCESS AND META-PROCESS. Everson Museum of Art. Videotapes and installations.

Organizations

Chicago. UNIVERSITY OF ILLINOIS AT CHICAGO. Dan Sandin and Tom DeFanti initiate video/computer graphics courses.
Minneapolis. UNIVERSITY COMMUNITY VIDEO. Center devoted to independent production, in 1981 begins exhibition and distribution.
New York. CABLE ARTS FOUNDATION. Founded by Russell Connor. Organization for production and distribution of anthology and art series to cable systems and for encouragement of local arts programming.
John Simon Guggenheim Foundation awards first video fellowship.
VISUAL RESOURCES. Director, Eva Kroy Wisbar. Distribution/information service including video. Publishes *Art & Cinema*, including coverage of video.
Portland, Ore. Northwest Film Study Center initiates Northwest Film and Video Festival. Directors include Robert Sitton and Bill Foster. In 1979 Film Study Center begins workshops and exhibitions in video.
Rochester, N.Y. VISUAL STUDIES WORKSHOP establishes media center. Production facility with workshops and exhibitions. Begins publication of *Afterimage* with coverage of video. Director, Nathan Lyons. Media center coordinators include Wayne Luke, Laddy Kite, Arthur Tsuchiya, Nancy Norwood.

Television/Productions

New York. Steve Rutt and Bill Etra develop Rutt/Etra scan processor.
San Francisco. VIDEOLA. San Francisco Museum of Art. Environmental sculpture by Don Hallock with multiple display of synthesized video works created at National Center for Experiments in Television (NCET), KQED-TV. Works by Stephen Beck with Don Hallock and Ann Turner, William Gwin with Warner Jepson, Don Hallock.

Publications

SPAGHETTI CITY VIDEO MANUAL by the Videofreex (New York: Praeger). Alternative equipment manual.

1974
Exhibitions/Events

Ithaca, N.Y. FIRST ANNUAL ITHACA VIDEO FESTIVAL, Ithaca Video Projects. In 1976 festival begins to tour.
Los Angeles. COLLECTOR'S VIDEO, Los Angeles County Museum of Art. Organizer, Jane Livingston. Works by John Baldessari, Peter Campus, Terry Fox, Frank Gillette, Nancy Holt, Joan Jonas, Paul Kos, Richard Landry, Andy Mann, Robert Morris, Bruce Nauman, Richard Serra, Keith Sonnier, William Wegman.
Minneapolis. NEW LEARNING SPACES AND PLACES, Walker Art Center. Includes installation by Frank Gillette and videotapes by James Byrne, Peter Campus, Juan Downey, Frank Gillette, Andy Mann, Ira Schneider, University Community Video, William Wegman.
PROJECTED IMAGES, Walker Art Center. Includes video installation by Peter Campus and performance with video with Joan Jonas.
New York. ELECTRONIC ART IV by Nam June Paik, Galeria Bonino.
OPEN CIRCUITS: THE FUTURE OF TELEVISION. The Museum of Modern Art. Organized by Fred Barzyk, Douglas Davis, Gerald O'Grady, Willard Van Dyke. International video conference with exhibition of tapes. Participants include museum educators and curators, cable and educational television producers, artists and art critics from U.S., Canada, Latin America, Europe, Japan.
PROJECTS: VIDEO, The Museum of Modern Art. Curator, Barbara London. Beginning of continuing series of video exhibitions. Program expands with funding from the Rockefeller Foundation in 1976.
VIDEO PERFORMANCE, 112 Greene Street, Video performances by Vito Acconci, Joseph Beuys, Chris Burden, Dennis Oppenheim, Ulrike Rosenbach, Richard Serra with Robert Bell, Willoughby Sharp, Keith Sonnier, William Wegman.
Syracuse. VIDEA 'N' VIDEOLOGY: NAM JUNE PAIK, 1959-73. Everson Museum of Art. Curator, David Ross. Retrospective of artist's videotapes, with catalog edited by Judson Rosebush.
VIDEO AND THE MUSEUM, Everson Museum of Art. Organized by David Ross. Funded by the Rockefeller Foundation. Conference with workshops for curators and administrators on the role of video in the museum. Concurrent, exhibitions: Peter Campus, *Closed Circuit Video*; Juan Downey, *Video Trans Americas Debriefing Pyramid* (a video/dance performance with Carmen Beuchat); Andy Mann, *Video Matrix*; and Ira Schneider, *Manhattan Is an Island.*
Washington, D.C. John F. Kennedy Center for the Performing Arts. Includes twenty-three videotapes.

Organizations

Bayville, N.Y. INTER-MEDIA ART CENTER (IMAC). Multipurpose production facility with post-production workshops and exhibitions. Director, Michael Rothbard.
Long Beach, Calif. f493LONG BEACH MUSEUM OF ART begins video exhibition program and collection of videotapes. Video curators include David Ross, Nancy Drew, Kathy Huffman. In 1976 production is moved to new facility and called the Station/Annex.

New York. ANTHOLOGY FILM ARCHIVES begins video program. Director, Jonas Mekas. Video Curators include Shigeko Kubota, Bob Harris. Includes exhibition, preservation, archive of videotapes and printed matter, screenings. In 1983 begins publication of *Video Texts*, an annual magazine on video art organized by Robert Haller, Bob Harris.
ASSOCIATION OF INDEPENDENT VIDEO AND FILMMAKERS (AIVF). Founded by Ed Lynch. Directors include Alan Jacobs, Lawrence Sapadin. National trade association of independent producers and individuals. Begins publishing *The Independent* on media issues. In 1975 establishes the Foundation for Independent Video and Film (FIVF) as an educational organization.
ANNA CANEPA VIDEO DISTRIBUTION (originally Video Distribution, Inc.). Distribution service of artists' tapes.
THE KITCHEN CENTER FOR VIDEO, MUSIC AND DANCE (formerly The Electronic Kitchen) relocates to Broome Street and begins daytime exhibition program. Inaugural show includes videotapes and three installations by Bill Viola.
Providence, R.I. ELECTRON MOVERS. Video art collective with gallery space, equipment resources, workshops, and visiting artist series. Founded by Dennis Hlynsky, Robert Jungels, Laurie McDonald, Alan Powell. In 1975 Ed Tannenbaum joins. Disbands 1980.
San Francisco. LA MAMELLE. Artists' space for video, audio, and marginal works. Directors, Carl Loeffler and Nancy Frank.
Seattle. AND/OR. Space for multimedia exhibitions, productions, performance art. In 1979 establishes 911, Video Library. In 1981 media program becomes Focal Point Media Center. Founded by Ann Focke, Robert Garner, Ken Leback, Video Curators, Norie Sato, Heather Oakson.

Television/Productions

Boston. NEW TELEVISION WORKSHOP, WGBH-TV. Established with grant from the Rockefeller Foundation and through the efforts of David Atwood, Fred Barzyk, Dorothy Chiesa, Ron Hays, Rich Hauser, Olivia Tappan. Director, Fred Barzyk. Producers include Dorothy Chiesa, Susan Dowling, Nancy Mason-Hauser, Olivia Tappan.
VIDEO: THE NEW WAVE, WGBH-TV. Program of video artists, including David Atwood, Stephen Beck, Peter Campus, Douglas Davis, Ed Emshwiller, Bill Etra, Frank Gillette, Don Hallock, Ron Hays, Nam June Paik, Otto Piene, Rudi Stern, Stan VanDerBeek, William Wegman. Writer and narrator, Brian O'Doherty.
New York. CUBA: THE PEOPLE by Jon Alpert and Keiko Tsuno, Public Broadcasting System (PBS). First documentary videotape using half-inch color equipment to be broadcast by public television.
Rochester, N.Y. TELEVISION WORKSHOP, WXXI-TV. Directors include Ron Hagell, Pat Faust, Carvin Eison. Ends 1981.

Publications

ARTS MAGAZINE (New York: Art Digest). Special video issue. Contributions by Eric Cameron, Russell Connor, Hermine Freed, Dan Graham, Shigeko Kubota, Bob and Ingrid Wiegand.
CYBERNETICS OF THE SACRED by Paul Ryan (Garden City, N.Y.: Anchor Press/Doubleday).

INDEPENDENT VIDEO, A COMPLETE GUIDE TO THE PHYSICS, OPERATION, AND APPLICATION OF THE NEW TELEVISION FOR THE STUDENT, ARTIST, AND FOR COMMUNITY TV by Ken Marsh (San Francisco: Straight Arrow Books).
THE PRIME TIME SURVEY by Top Value Television (TVTV). Report on status of video and its directions.

1975

Exhibitions/Events

Dallas. THE ETERNAL FRAME by T. R. Uthco and Ant Farm. Reenactment of John F. Kennedy assassination for videotape. Presented as installation at Long Beach Museum of Art in 1976.
Long Beach, Calif. SOUTHLAND VIDEO ANTHOLOGY. Long Beach Museum of Art. Extended series of five exhibitions by California artists.
AMERICANS IN FLORENCE, EUROPEANS IN FLORENCE, Long Beach Museum of Art. Organized by Maria Gloria Bicocchi and David Ross. Traveling exhibition with videotapes produced by Art/Tapes/22, Florence.
New York. FIRST ANNUAL VIDEO DOCUMENTARY FESTIVAL, initiated by Video Study Center of Global Village.
1975 BIENNIAL EXHIBITION, Whitney Museum of American Art. Includes work by eighteen video artists.
PROJECTED VIDEO, Whitney Museum of American Art. Projected videotapes by William Adler and John Margolies, John Baldessari, Lynda Benglis, Peter Campus, Douglas Davis, Bill Etra, Hermine Freed, Shigeko Kubota, Nam June Paik, Richard Serra, Keith Sonnier, Steina and Woody Vasulka, William Wegman.
Philadelphia. VIDEO ART, Institute of Contemporary Art, University of Pennsylvania. Curator, Suzanne Delehanty. Exhibition documenting the development of video art through videotapes and installations. Travels to Contemporary Art Center, Cincinnati; Museum of Contemporary Art, Chicago; Wadsworth Atheneum, Hartford, Conn.; and Sao Paulo Biennale, Sao Paulo, Brazil.
San Francisco. MEDIA BURN by Ant Farm, Cow Palace. July Fourth performance/media event.
MOEBIUS VIDEO SHOW, San Francisco Art Festival. First exhibition of video in the Art Festival. Includes work by Ant Farm, Terry Fox, Phil Garner, Joanne Kelly, Darryl Sapien, Skip Sweeney.
WALK SERIES by Peter D'Agostino, 80 Langton Street. Video installation and first event at 80 Langton Street, an alternative space initially sponsored by the San Francisco Art Dealers Association. In 1976 becomes an independent space with emphasis on alternative art forms.

Organizations

Hartford, Conn. REAL ART WAYS. Arts center with video exhibitions and library. Video coordinators include David Donihue, Gary Hogan, Ruth Miller.
New York. INDEPENDENT CINEMA ARTISTS AND PRODUCERS (ICAP) forms to represent independent film and video artists to cable systems. President, Kitty Morgan.

Television/Productions

New York. VIDEO AND TELEVISION REVIEW (VIR), the Television Laboratory at WNET/Thirteen. Executive Producer, Carol Brandenburg. Yearly broadcast series of tapes from U.S. and Europe. In 1979 renamed Video/Film Review.

1976

Exhibitions/Events

Berkeley, Calif. COMMISSIONED VIDEO WORKS, University Art Museum. Organized by Jim Melchert. Fifteen artists commissioned to make tapes of under four-minute duration. Includes Eleanor Antin, David Askevold, Siah Armajani, John Baldessari, Robert Cumming, John Fernie, Hilla Futterman, Leonard Hunter, Anda Korsts, Les Levine, Paul McCarthy, George Miller, Dennis Oppenheim, Robert Watts, William Wegman.
Boston. CHANGING CHANNELS, Museum of Fine Arts and Museum School Gallery. Exhibition of videotapes produced by independent artists at experimental television broadcast centers: WGBH, Boston; WNET, New York; and KQED, San Francisco.
San Francisco. VIDEO ART: AN OVERVIEW, San Francisco Museum of Modern Art. Organized by David Ross. Exhibition of thirty-three videotapes by twenty-nine artists. Installations by Peter Campus, Paul and Marlene Kos, Nam June Paik.
Syracuse, N.Y. NEW WORK IN ABSTRACT VIDEO IMAGERY. Everson Museum of Art. Curator, Richard Simmons. Works by forty artists using synthesizers, lasers, and computers.

Organizations

Boston. BOSTON FILM/VIDEO FOUNDATION. Offers screenings, educational programs, equipment resources. Founded by Jon Rubin and Susan Woll. Direc tors include Michelle Schofield and Tom Wylie.
Chicago. VIDEO DATA BANK. School of the Art Institute of Chicago. Distribution and resource center for videotapes on artists and video art. Director, Lyn Blumenthal.
New York. ASIAN CINE-VISION. Media center in Chinatown producing Asian-American program series and programming for Chinese Cable Television. Conducts workshops, media and production services, and operates an Asian-American Media Archive. In 1982 begins Asian-American International Video Festival. Director, Peter Chow.
DONNELL LIBRARY CENTER, New York Public Library, establishes collection of videotapes. Founded by William Sloan. Video librarians have included Mary Feldstein, Michael Miller, Michael Gitlin, Lishin Yu.
FRANKLIN FURNACE. Alternative space with archive, bibliography, exhibition, performance programs, including video. Director, Martha Wilson.
NEW AMERICAN FILMMAKER SERIES, Whitney Museum of American Art. Continuing exhibition of independent film expands to include video art. Director, John Hanhardt.
Pittsburgh. INDEPENDENT FILM AND VIDEO PREVIEW NETWORK. Pitts-

burgh Filmmakers. Program of organized preview screenings of films and videotapes around the country. Founded by Sally Dixon and Robert Haller. Ends 1980.
San Francisco. BAY AREA VIDEO COALITION founded with grant from the Rockefeller Foundation. Production/post-production center with workshops and exhibitions. Founding Director, Gail Waldron. Director, Morrie Warshawski.

Television/Productions

Los Angeles. VIDEO ART. Los Angeles Theta Cable, Long Beach Cablevision, and Santa Barbara Cable TV. Cable series produced by Some Serious Business and the Long Beach Museum of Art. Ends 1979.
New York. CABLE SOHO. President, Jaime Davidovich. Independent organization for innovative arts programming on cable television. In 1977 becomes Artists' Television Network.
IMAGE UNION. Independent production company forms to offer alternative coverage of the Democratic National Convention and Election Night. *The Five-Day Bicycle Race* and *Mock Turtle Soup*, taped segments with live phone-in interviews, are shown on Manhattan Cable Television.

Television/Productions

VIDEO ART: AN ANTHOLOGY (New York: Harcourt, Brace and Jovanovich). Editors, Beryl Korot and Ira Schneider. First anthology of video criticism and statements by video artists.
VIDEO: STATE OF THE ART by Joanna Gill (New York: The Rockefeller Foundation). Report on video activity in the United States.

1977

Organizations

Atlanta. IMAGE FILM/VIDEO CENTER (Independent Media Artists of Georgia, Etc., Inc.). Media center with screenings, workshops, and equipment access. Begins the Atlanta Independent Film and Video Festival (now the Atlanta Film and Video Festival), an annual international showcase. Directors include Gayla Jamison, Anna Marie Piersimoni, Marsha Rifkin.
Houston. SOUTHWEST ALTERNATIVE MEDIA PROJECT (SWAMP). Originally associated with the Rice Media Center at Rice University. Media center with education program, lecture series, production and post-production technical assistance. Conducts Southwest Film and Video Tour, artist-in-residence program, and annual Texpo film and video festival. Produces local PBS series, "The Territory." Directors include Ed Hugetz and Tom Sims.
New York. LOCUS COMMUNICATIONS. Equipment access center with workshops, technical production services, cable programming, screenings. Founding Executive Director, Gerry Pallor.
Port Washington, N.Y. PORT WASHINGTON LIBRARY begins visiting artists program with exhibitions and presentations. Head of Media Services, Lillian Katz.

Television/Productions

Buffalo, N.Y. Steina and Woody Vasulka and Jeffrey Schier begin work on the Digital Image Articulator, a digital computer-imaging device.
Chicago. ZGRASS. Personal computer-graphics system designed by artist Tom DeFanti.
Los Angeles. THE SATELLITE ARTS PROJECT by Kit Galloway and Sherrie Rabinowitz. Live interactive broadcast between California, Maryland, and Washington, D.C.
New York. DOCUMENTA VI. Curator, Wulf Herzogenrath. Satellite performance project with Joseph Beuys, Douglas Davis, and Nam June Paik broadcast internationally from Kassel, West Germany, presented through WNET-TV.
INDEPENDENT DOCUMENTARY FUND, WNET-TV. Executive Producer, David Loxton. Coordinator, Kathy Kline. Established at the Television Laboratory with grants from the Ford Foundation and the National Endowment for the Arts to stimulate the production of independent documentaries.
New York and San Francisco. SEND/RECEIVE SATELLITE NETWORK. Coordinators Liza Bear and Keith Sonnier with support from the Public Interest Satellite Association (PISA) and NASA. Two-way satellite transmission between New York and San Francisco with simultaneous performances. Participants, in San Francisco: Margaret Fischer, Terry Fox, Brad Gibbs, Sharon Grace, Carl Loeffler, Richard Lowenberg, Alan Scarritt. In New York: Liza Bear, Richard Landry, Nancy Lewis, Richard Peck, Betsy Sussler, Willoughby Sharp, Paul Shavelson, Duff Schweiniger, Keith Sonnier.

Publications

THE NEW TELEVISION: A PUBLIC/PRIVATE ART. (Cambridge, Mass. and London: The MIT Press). Manifesto including essays from the Open Circuits Conference at The Museum of Modern Art, New York, in 1974.

1978

Exhibitions/Events

Buffalo. VASULKA: STEINA-MACHINE VISION, WOODY-DESCRIPTION, Albright-Knox Gallery. Curator, Linda L. Cathcart. Exhibition of tapes and installations.
New York. ARANSAS, AXIS OF OBSERVATION by Frank Gillette, The Kitchen. Travels to Contemporary Arts Museum, Houston; University Art Museum, Berkeley; and Academy of Fine Arts, Washington, D.C. Acquisitioned by University Art Museum.
VIDEO VIEWPOINTS. The Museum of Modern Art. Beginning of yearly lecture series by independent videomakers.
Pittsburgh. NATIONAL MEDIA ALLIANCE OF MEDIA ARTS CENTERS (NAMAC) holds first conference. Hosted by Pittsburgh Filmmakers.
Redington Beach, Fla. CHINSEGUT FILM/VIDEO CONFERENCE. Founded by Charles Lyman and Peter Melaragno. Conference with presentations to promote interchange among invited participants and film- and video-makers.
Venice, Calif. VIDEO NIGHT by Some Serious Business. Weekly video screening series.

Organizations

Chicago. CHICAGO EDITING CENTER. Production/post-production facility with education and exhibition programs. In 1980 becomes Center for New Television. Directors include Cynthia Neal, Joyce Bollinger.

Organizations

Chicago. IMAGE UNION, WTTW-TV. Produced by Tom Weinberg. Weekly broadcast of independent work.
New York. ARTISTS' TELEVISION NETWORK initiates "Soho Television," regular programming of artists' videotapes and performances, and of "The Live! Show," avant-garde variety show. Director, Jaime Davidovitch.
POTATO WELL. Collaborative Projects. Artists' television series for cable begins as live show and evolves into diversified programming with emphasis on narrative and performance-oriented work involving artists from diverse media. Regular producers include Cara Brownell, Mitch Corber, Albert Dimartino, Julie Harrison, Robert Klein, Terry Mohre, Alan Moore, Brian Piersol, Gary Pollard, Mindy Stevenson, Jim Sutcliffe, Maria Thompson, Sally White.

1979

Exhibitions/Events

Long Beach, Calif. N/A VISION, sponsored by Long Beach Museum of Art. Weekly circulating video screening series at Long Beach Museum of Art, Foundation of Art and Resources (FAR), and Highlands Art Agents.
New York. RE-VISIONS: PROJECTS AND PROPOSALS IN FILM AND VIDEO, Whitney Museum of American Art. Curator, John Hanhardt. Video installations by Bill Veirne; David Behrman, Bob Diamond and Robert Watts; and Buky Schwartz.
VIDEOTAPES BY BRITISH ARTISTS, The Kitchen. Curator, Steve Partridge. Works by David Crichley, David Hall, Tamara Krikorian, Stuart Marshall, Steve Partridge, and others.
VIDEO FROM TOKYO TO KUKUI AND KYOTO. The Museum of Modern Art. Curator, Barbara London. A survey of the works of thirteen contemporary Japanese artists. Travels to Long Beach Museum of Art, Long Beach, Calif.; Vancouver Art Gallery, Vancouver, B.C.; and with "Video New York, Seattle and Los Angeles" travels to Japan and Europe.
Syracuse, N.Y. EVERSON VIDEO REVUE. Everson Museum of Art. Curator, Richard Simmons. Exhibition with videotapes by over fifty artists. Travels to Museum of Contemporary Art, Chicago; University Art Museum, Berkeley, Ca.; in 1981, Museum of Contemporary Art, La Jolla, Ca.
Berkeley, Calif. UNIVERSITY ART MUSEUM, University of California at Berkeley institutes regular weekend programming. Organized by David Ross. Ends 1981.
New York. THE MEDIA ALLIANCE. Association of media arts organizations and independent video producers in New York State designed to coordinate resources and promote the work of the independent video community. Includes program-

ming, exhibition, production, distribution. Directors include Jackie Kain, Robin White.
P.S. 1 begins video exhibition program with emphasis on installations. Video Curator, Bob Harris.

Television/Productions

New York. COMMUNICATIONS UPDATE. Center for New Art Activities. Originally the WARC (World Administrative Radio Conference) Report. Artists series for cable dealing with political and communications issues. Original producers: Liza Bear, Rolf Brand, Michael McClard, Willoughby Sharp. In 1983 becomes Cast Iron TV and programming diversifies. Producer, Liza Bear.
NON-FICTION TELEVISION, WNET/Thirteen. Broadcast series for Independent Documentary Fund.
PUBLIC INTEREST VIDEO NETWORK. Executive Producer, Kim Spencer. Senior Editor, Nick DeMartino. Independent production company financed by the Urban Scientific and Educational Research (USER) presents live satellite coverage of an antinuclear demonstration in Washington, D.C., on the Public Broadcasting System (PBS). First time PBS carries a live public affairs program whose editorial content was determined by an organization outside its system.
San Francisco. PRODUCED FOR TELEVISION, La Mamelle and KTSF-TV. Live broadcast of performance art. Works by Chris Burden, Lynn Hershman and Rea Baldridge, Chip Lord and Phil Garner, Barbara Smith.

Publications

VIDEO-ARCHITECTURE-TELEVISION: WRITING ON VIDEO AND VIDEO WORKS by Dan Graham (Halifax, Nova Scotia and New York: The Press of the Nova Scotia College of Art and Design and the New York University Press).

1980

Exhibitions/Events

Berkeley, Calif. and New York. VIDEO ABOUT VIDEO: FOUR FRENCH ARTISTS, University Art Museum, University of California; and Télétheque-Alliance Francaise, New York. Works by Paul-Armaud Gette, Philippe Oudard, Philippe Guerrier, Thierry Kuntzel.
Buffalo, N.Y. INSTALLATION: VIDEO, Hallwalls. Exhibition with work by Dara Birnbaum, Patrick Clancy, Wendy Clarke, Brian Eno, Ken Feingold, Dan Graham, Gary Hill, Sarah Hornbacher, Shigeko Kubota.
Lake Placid, N.Y. ART AT THE OLYMPICS, 1980 WINTER GAMES. Videotapes by Skip Blumberg, Kit Fitzgerald and John Sanborn, Nam June Paik. Installations by Wendy Clarke, Frank Gillette, Ira Schneider, Buky Schwartz.
Long Beach, Calif. CALIFORNIA VIDEO, Long Beach Museum of Art. Curator, Kathy Huffman. Works by Max Almy, Dan Boord, Ante Boznich, John Caldwell, Alba Cane, Helen DeMichiel, Tony Labat, Pier Marton, Tony Oursler, Jan Peacock, Patti Podesta, Joe Rees/Target Video, Nina Salerno, Ilene Segalove, Starr Sutherland, "Captain" Bruce Walker, Bruce and Norman Yonemoto.
New York. LOVE TAPES IN NEW YORK by Wendy Clarke. Live interactive

installation and tapes exhibited at the World Trade Center with selections shown on cable television and WNET/Thirteen.
TELEVISION/SOCIETY/ART. The Kitchen. Organized by Ron Clark and Mary MacArthur. Colloquium presented by The Kitchen and the American Film Institute. Participants include Benjamin Buchloh, Julianne Burton, Nick DeMartino, Stephen Heath, Fredric Jameson, Rosalind Krauss, Mark Nash, Robert Sklar, Martha Rosler, Herbert Schiller, Allan Sekula, Peter Wollen.
San Francisco. FIRST ANNUAL SAN FRANCISCO VIDEO FESTIVAL. Director, Steve Agetstem. Assistant Director, Wendy Garfield. Begin publishing *Video 80* as festival catalog. Now called *SEND* and published as a quarterly.
Yonkers, N.Y. ALTERNATIVE SPACES. Hudson River Museum. Series of exhibitions employing Museum's planetarium. Includes video installations by Mary Lucier, Francesc Torres.

Organizations

New Orleans. SURVIVAL INFORMATION TELEVISION, NOVAC. Installation in local Welfare Office with social issues programming run on a repeating cycle.
St. Paul. Jerome Foundation expands to award grants to video artists.

Television

Cambridge. ARTISTS' USE OF TELECOMMUNICATIONS. Organized by Center for Advanced Visual Studies, Massachusetts Institute of Technology (MIT). Collaborative interactive slow-scan TV conference link between Cambridge, New York, San Francisco, Long Beach, Toronto, Vienna, Tokyo, and Vancouver.
THREE ARTISTS ON LINE IN THREE COUNTRIES. Three-way slow-scan transmission between Aldo Tambellini, Cambridge, Tom Klinkowstein, Amsterdam, and Bill Bartlett, Vancouver.
Los Angeles and New York. HOLE-IN-SPACE by Kit Galloway and Sherrie Rabinowitz. Live interactive satellite project between Los Angeles and New York.
Minneapolis-St. Paul. MINNESOTA LANDSCAPES. KTCA-TV. Project Director, Peter Bradley. Series of videotapes on Minnesota for broadcast. Works by Skip Blumberg, James Byrne, Steve Christiansen, Davidson Gigliotti, Frank Gohlke, Cynthia Neal, Steina.

1981

Exhibitions/Events

New York. FIRST NATIONAL LATIN FILM AND VIDEO FESTIVAL, El Museo del Barrio.
1981 BIENNIAL EXHIBITION, Whitney Museum of American Art. Installations by Frank Gillette and Buky Schwartz.
STAYED TUNED, The New Museum. Organized by Ned Rifkin. Exhibition juxtaposes artists' work in video with work in other media. Includes Robert Cumming, Brian Eno, Charles Frazier, Donald Lipski, Howardena Pindell, Judy Rifka, Allen Ruppersberg, Irvin Tepper.
VIDEO CLASSICS. Bronx Museum of the Arts. Curator, RoseLee Goldberg. Installations by Vito Acconci, Dan Graham, Shigeko Kubota, Rita Myers, Bruce Nauman, Dennis Oppenheim, Nam June Paik.

Rochester, N.Y. FROM THE ACADEMY TO THE AVANT-GARDE. Visual Studies Workshop. Curator, Richard Simmons. Traveling exhibition with videotapes by Juan Downey, Howard Fried, Frank Gillette, Davidson Gigliotti, Tony Labat, Les Levine. Travels to Center for Art Tapes, Halifax, Nova Scotia, and Center for New Television, Chicago.
Washington, D.C. NATIONAL VIDEO FESTIVAL, American Film Institute. Sponsor, Sony Corporation. Festival producer, Larry Kirkman; festival director, James Hindman. Installation by Nam June Paik.

Organizations

Pittsburgh. MUSEUM OF ART, Carnegie Institute, expands its Film Section to the Section of Film and Video, and opens Video Gallery. Curator of Film and Video, William Judson.

Television/Productions

New York and Paris. DOUBLE ENTENDRE by Douglas Davis, Whitney Museum of American Art and Centre Georges Pompidou, Paris. Satellite telecast perfor mance.
New York. PAPER TIGER TELEVISION. Organized by Diane Augusta, Pennee Bender, Skip Blumberg, Shulae Chang, DeeDee Halleck, Cayn Rogoff, David Shulman, Alan Steinheimer. Series on public-access television that examines communications industry via the print media, and serves as model for low-budget, public-access programming.

1982

Exhibitions/Events

Boston. SIGGRAPH (Special Interest Group in Computer Graphics). Annual conference includes computer-generated video art in its juried art show. Organized by Copper Giloth.
Buffalo, N.Y. ERSATZ TV: A STUDIO MELEE by Alan Moore and Terry Mohre, Collaborative Projects. Hallwalls Gallery. Curator, Kathy High. Installations of six studio sets from artists' television series "Potato Wolf," with live cameras and videotape screenings.
VIDEO/TV: HUMOR/COMEDY, Media Study/Buffalo. Curator, John Minkowsky. Touring exhibition that explores relationship between art and entertainment. Travels throughout U.S.
New York. NAM JUNE PAIK, Whitney Museum of American Art. Director of exhibition, John Hanhardt. Major retrospective. Travels to Museum of Contemporary Art, Chicago.
Park City, Utah. FOURTH ANNUAL UNITED STATES FILM AND VIDEO FESTIVAL expands to include video.
Yonkers, N.Y. ART AND TECHNOLOGY: APPROACHES TO VIDEO, Hudson River Museum. Three-part exhibition of installations by Dara Birnbaum, David Behrman and Paul DeMarinis, and Kit Fitzgerald and John Sanborn. Curator, Nancy Hoyt.

Washington, D.C. NATIONAL VIDEO FESTIVAL, American Film Institute at the John F. Kennedy Center for the Performing Arts, and the American Film Institute Campus, Los Angeles. Sponsor, Sony Corporation. Installations by Shigeko Kubota (Washington, D.C.) and Ed Emshwiller and Bill Viola (Los Angeles).

Organizations

Boston. INSTITUTE OF CONTEMPORARY ART begins video program. Director, David Ross.

Portland, Ore. THE MEDIA PROJECT. Expands to include video. Media organization for distribution of independent work includes workshops and state-wide directory of media services, and acts as a liaison to cable. Director, Karen Wickery.

Television/Productions

Los Angeles. THE ARTIST AND TELEVISION: A DIALOGUE BETWEEN THE FINE ARTS AND THE MASS MEDIA. Sponsored by ASCN Cable Network, Los Angeles, and University of Iowa, Iowa City. Interactive satellite telecast connecting artists, critics, curators, and educators in Los Angeles, Iowa City, and New York.

New York. DISARMAMENT VIDEO SURVEY. Organized by Skip Blumberg, Wendy Clarke, DeeDee Halleck, Karen Ranucci, Sandy Tolan. Collaboration by over 300 independent producers from New York, Washington, D.C., San Francisco, Great Britain, Germany, Japan, India, the Netherlands, Mexico, Brazil, and other locations to compile one-minute interviews with people about their views on nuclear arms and disarmament. Survey shown on cable television and presented as installations at American Film Institute National Video Festival in Washington, D.C.

THE VIDEO ARTIST, Producers: Eric Trigg, Electronic Arts Intermix, Stuart Shapiro. Sixteen-part series on major video artists broadcast nationally over USA Cable Network.

1983

Exhibitions/Events

Minneapolis. THE MEDIA ARTS IN TRANSITION. Conference organizers and sponsors: Walker Art Center, National Alliance of Media Arts Centers (NAMAC), Minneapolis College of Art and Design, University Community Video, Film in the Cities. Conference programmers: Jennifer Lawson, John Minkowsky, Melinda Ward.

New York. THE INTERSECTION OF THE WORD AND THE VISUAL IMAGE, Women's Interart Center. Colloquium involving artists, writers, and scholars on relationship of language to the moving image, alternative narratives, and the transformation of literary, historical, performance, and visual works to video. Screenings of international works.

1983 BIENNIAL EXHIBITION, Whitney Museum of American Art. Installations by Shigeko Kubota and Mary Lucier. First touring video show of Biennial, through American Federation of Arts (AFA).

Rochester, N.Y. VIDEO INSTALLATION 1983, Visual Studies Workshop. Exhibition including works by Barbara Buckner, Tony Conrad, Doug Hall, Margia Kramer, Bill Stephens.

Sante Fe and Albuquerque. VIDEO AS ATTITUDE, Museum of Fine Arts, Santa Fe, and University Art Museum, Albuquerque, New Mexico. Director, Patrick Clancy. Installations by Bill Bierne, Juan Downey, Dieter Froese, Robert Gaylor, Gary Hill, Joan Jonas, Rita Myers, Bruce Nauman, Michael Smith, Steina, Francesc Torres, Bill Viola.

Valencia, Calif. HAJJ by Mabou Mines, California Institute of the Arts. Written by Lee Brerer, performed by Ruth Meleszech. Video by Craig Jones. Premiere performance of complete version of performance poem, which incorporates extensive use of live and recorded videotape.

Yonkers, N.Y. ELECTRONIC VISION, Hudson River Museum. Curator, John Minkowsky. Installations by Gary Hill, Ralph Hocking and Sherry Miller, Dan Sandin, Steina and Woody Vasulka.

New York and Long Beach, Calif. THE SECOND LINK: VIEWPOINTS ON VIDEO IN THE EIGHTIES. Organized by Lorne Falk, Walter Phillips Gallery at the Banff Centre School of Fine Arts. United States showing at The Museum of Modern Art and Long Beach Museum of Art. Curators, Peggy Gale, Kathy Huffman, Barbara London, Brian McNevin, Dorine Mignot, Sandy Nairne. Works from Europe, Canada. U.S. International tour.

Television/Productions

Long Beach, Calif. SHARED REALITIES. Long Beach Museum of Art. Executive Producer, Kathy Huffman. Series on local cable station of work produced by artists at the Station/Annex, programming about the museum, and local cultural programming.

New York. PERFECT LIVES by Robert Ashley. Project Director, Carlota Schoolman. Video Director, John Sanborn. Television opera in seven parts produced by The Kitchen.

Minneapolis/St. Paul. ALIVE FROM OFF CENTER. KTCA-TV and Walker Arts Center begin production of a series of programs bringing together performance artists and television. Melinda Ward, Executive Producer. Acquisitions and new productions scheduled for 1985 broadcast.

1984

Exhibitions/Events

Chicago. VIDEO DRIVE-IN. The Video Data Bank. Outdoor evening video festival in Grant Park. Includes independent and commercial work and a multi-media event directed by Ed Paschke.

Long Beach, Calif. VIDEO: A RETROSPECTIVE. Parts I and II. Long Beach Museum of Art. Connie Fitzsimons and Kira Perov, curators. Installations by Shigeko Kubota, Celia Shapiro, John Sturgeon, Charlemagne Palestine, and performance by Terry Fox.

Los Angeles, Calif. THE NATIONAL VIDEO FESTIVAL OLYMPIC SCREENINGS. Organized by the American Film Institute. Sponsored by the Olympic Arts Festival. Executive Director, James Hindman. Director, Jacqueline Kain. Curators include Kathy Huffman, Long Beach Museum of Art; Julie Lazar, Museum of Contemporary Art; Mark Holmes, Los Angeles Institute of Contemporary Art; VIDEOLACE, Los Angeles Contemporary Exhibitions. Features video explorations of dance, art, music, performance, and television, reflecting the scope of the Olympic Arts Festival.

San Francisco, Calif. LORNA, by Lynn Hershman. Fuller Goldeen Gallery. First artist's project to incorporate interactive laser disc technology. Production begun for "TV on TV" Texas Tech, Lubbock, Texas. Produced by the Electronic Arts Archive, Texas Tech.

New York. VIDEO: A HISTORY. Parts I and II. The Museum of Modern Art. Directed by Barbara London. Exhibition traces the evolution of video art from its beginning in the early 1960s through the present through photos, texts, publications, and videotapes.

New York/Paris. GOOD MORNING MR. ORWELL. Organized by Nam June Paik. Co-production of WNET/THIRTEEN Television Laboratory, New York: FR3 (French National Television), Centre Georges Pompidou, Paris; and WDR3, Cologne. An interactive satellite project between New York and Paris with live performances by John Cage, Merce Cunningham, Allen Ginsverg, Mitchell Kriegman, Charlotte Moorman, Peter Orlovsky, Joseph Beuys, Robert Combas, Pierre-Alain Hubert, Sapho, Studio Bercot, Urban Sax, and Ben Vauthier. Hosts, George Plimpton, Jacques Villers.

Lubbock, Texas. TV ON TV. Carl Loeffler and Kim Smith, co-executive producers. Sponsored by Electronic Arts Archive and Research Institute, Texas Tech University. Works produced on location at Lubbock, designed for broadcast on television. Artists include Michael Smith, Bruce and Norman Yonemoto, Jaime Davidovich, Marsella Bienvenue, and Lynn Hershman.

Organizations

Boston. CONTEMPORARY ART TELEVISION FUND. (CAT Fund). Collaboration of WGBH-TV and Institute oof Contemporary Art for the support of production of new works using television. Kathy Huffman, Curator. First co-productions by Joan Jonas, Bill Seaman, Tony Oursler, Chip Lord and Mickey McGowan. Organizes one-day symposium on the distribution of artist's video, with curaators, artists, and distributors from the United States, Canada, and Europe. Michele Schofield, symposium coordinator.

Polaroid Corporation grants First Polaroid Video Art Award to Bill Viola.

Publications

VIDEO: A RETROSPECTIVE 1974-1984. Long Beach Museum of Art. Kathy Huffman, Editor. Managing Editors, Kira Perov and Connie Fitzsimons. Published in conjunction with "Video: A Retrospective" exhibition.

* * *

Videography

About Media. 25 min. color video, 1977. Prod: Tony Ramos. Dist: EAI.

An American Family. 12 60 min. programs. color video, 1972 Prod: Craig Gilbert at WNET. Dist: PBS, Washington DC.

The Best of William Wegman. 20 min. b&w video 1970-78. Prod: William Wegman. Dist: EAI/Castelli-Sonnabend.

The Bad Sister. 85 min. color video, 1983. Prod: Laura Mulvey and Peter Wollen for Channel Four. Dist: The Kitchen.

Chinatown. 50 min. color video, 1976. Prod: Jon Alpert and Keiko Tsumo for Downtown Community Television. Dist: EAI.

Eugene. 30 min. b&w kinescope, 1957 Prod: Ernie Kovacs at NBC. Dist: Excerpts are available on *Ernie Kovacs: Television's Original Genius* from Vestron Video, Stamford, CT.

Four More Years. 60 min. b&w video 1972. Prod: TVTV (Top Value Television: Michael Shamberg, Ira Schneider, Paul Ryan, Wendy Apple, Skip Blumberg, Nancy Cain and others.) Dist: EAI

Ghost Trio. 60 min. b&w film to video 1976. Prod: Samuel Beckett at BBC. Dist: BBC, London.

Global Groove. 30 min. color video 1973. Prod: Nam June Paik at WNET. Dist: EAI.

Hiroshima-Nagasaki, August 1945. 16 min. b&w film 1970. Prod: Erik Barnouw. Dist: MoMA. (EAI for video version.)

The Irish Tapes. 46 min. b&w video 1974. Prod: John Reilly and Stefan Moore. Dist: EAI/MoMA.

It Starts at Home. 25 min. color video 1982. Prod: Michael Smith. Dist: Castelli-Sonnabend.

The Laughing Alligator. 28 min. color video 1979. Prod: Juan Downey at WNET. Dist: Castelli-Sonnabend.

Media Burn. 25 min. 1975. Prod: Ant Farm (Chip Lord, Doug Michels and Curtis Schrier). Dist: EAI/MoMA.

Media Ecology Ads. 14 min. color video 1982. Prod: Antonio Muntadas. Dist: EAI/Video Data Bank.

The Medium is the Medium. 30 min. b&w and color video 1969. Prod: Fred Barzyk at WGBH. Includes: "Black" by Aldo Tambellini "Electronic Light Ballet" by Otto Piene, "Hello" by Alan Kaprow, and other segments. Dist: EAI.

Meta Mayan II. 20 min. color video 1981. Prod: Edin Velez and Ethel Velez at WNET. Dist: EAI/MoMA.

Music Word Fire and I Would Do it Again: The Lessions. 28 min. color video 1981. Prod: Robert Ashley with John Sanborn and Kit Fitzgerald at WNET. Dist: Kitchen/MoMA.

My Father Sold Studebakers. Prod: Skip Sweeney. Dist: Video Free America, San Francsico, CA.

Nova: A Whisper from Space. 60 min. color video 1978. Prod: Peter Jones, BBC/WGBH. Dist: King Features Syndicate, New York, NY.

O Superman. 8 min. color video, 1981. Prod: Laurie Anderson. Dist: Warner Bros. Records, Burbank, CA.

Odyssey: Franz Boas(1858-1942) 60 min. color video, 1980. Prod: Michael J. Ambrosino. Dist: Public Broadcasting Associates, Newton, MA.

Paper Tiger Television. 30 min. continuing cable-TV series—color video. Prod: Dee Dee Halleck, Skip Blumberg, Diana Agosta, Esti Marpet and others. Includes programs with Herb Schiller, Myrna Bain, Joan Braderman, Brian Winston, Martha Rosler, and others. Dist: Paper Tiger/Video Data Bank.

Perfect Leader. 4 min. color video, 1983. Prod: Max Almy. Dist: EAI.

Pick Up Your Feet: The Double Dutch Show. 30 min. color video, 1981. Prod: Skip Blumberg at WNET. Dist: EAI/MoMA.

Poet at Large: A Conversation with Robert Bly. 60 min. color video 1979. Prod: Bill Moyers' Journal at WNET. Dist: WNET.

The Police Tapes. 90 min. b&w video 1977. Prod: Alan and Susan Raymond at WNET. Dist: The Kitchen.

Reverse Television. 43 30 second spots, color video, 1983. Prod: Bill Viola at WGBH. Dist: EAI.

Savage/Love. 26 min. color video. Prod: Shirley Clark. Dist: Women's Interart Center, New York, NY

Scapemates. 29 min. color video 1972. Prod: Ed Emshwiller at WNET. Dist: EAI.

Shock of the New. 8 52 min. programs, color video, 1980. Prod: Robert Hughes. Dist: Time/Life Video, New York, NY

Smothering Dreams. 23 min. color video 1981. Prod: Dan Reeves with Jon Hilton at WNET. Dist: EAI/MoMA.

Technology/Transformation: Wonder Woman. 7 min. color video 1979. Prod: Dara Birnbaum. Dist: EAI/Video Data Bank.

TeleTapes. 28 min. color video 1981. Prod: Peter D'Agostino at WNET. Dist: EAI/MoMA.

Television Delivers People. 6 min. color video. Prod: Richard Serra with Carlota Fay Schoolman. Dist: Castelli-Sonnabend/EAI.

Three Pieces for Television Broadcast. 4 min. color video 1976. Prod: Chris Burden. Dist: Ronald Feldman Fine Arts, N.Y.

Upside Down and Backwards. 28 min. color video 1979. Prod: Joan Jonas. Dist: EAI.

Video Against Video. 30 min. color video 1975. Prod: Douglas Davis. Dist: EAI.

Vietnam: A Television History. 13 60 min. programs, color video and film, 1983. Prod: Stanley Karnouw at WGBH. Dist: Films, Inc.

The War Game. b&w 16mm film 1965. Prod: Peter Watkins at BBC. Dist: Films, Inc.

Ways of Seeing. 4 25 min. color 16mm film & video. 1974. Prod: John Berger and Michael Dibbs at BBC. Dist: Films Inc.

Why I got into TV and other stories. 10 min. color video. Prod: Ilene Segalove. Dist: AFA.

Distribution Catalogues

The American Federation of the Arts 41 East 65th St. New York, NY 10021

Castelli-Sonnabend Videotapes and Films 142 Greene St. New York, NY 10012

Electronic Arts Intermix 10 Waverly Pl. New York, NY 10003

Films Incorporated 8124 N. Central Park Ave. Skokie, IL 66007

The Kitchen/Video Distribution 59 Wooster Street New York, NY 10012

The Museum of Modern Art Circulating Video Library 11 W. 53rd St. New York, NY 10019

Video Data Bank, School of the Art Institute of Chicago Columbus Drive at Jackson Blvd. Chicago, IL 60603

Bibliography

Adler, Richard. Ed. *Understanding Television: Television as a Social and Cultural Force*. New York: Praeger Books, 1981.

Adorno, Theodor and Horkheimer, Max. *Dialectic of Enlightenment*. New York: Seabury Press, 1972.

Arendt, Hannah. *The Human Condition*. New York: Doubleday/Anchor Press, 1959.

Arlen, Michael. *The Living Room War*. New York: Viking Press, 1969.

____________. *The Camera Age*. New York: Farrar, Straus and Giroux, 1981.

Baggaley, Jon and Duck, Steven. *Dynamics of Television*. London: Gower Press, 1976.

Barnouw, Erik. *The Golden Web: History of Broadcasting in the United States*. New York: Oxford University Press, 1968.

____________. *Tube of Plenty: The Evolution of American Television*. New York: Oxford University Press, 1975.

____________. *The Sponsor: Notes on a Modern Potentate*. New York: Oxford University Press, 1978.

Barthes, Roland. *Image/Music/Text*. New York: Hill and Wang, 1977.

____________. *Mythologies*. Trans. Annette Lavers New York: Hill and Wang, 1977.

____________. *S/Z*. Trans. Richard Howard. New York: Hill and Wang, 1974.

Bateson, Gregory. *Steps to an Ecology of Mind*. New York: Random House, 1972.

Bateson, Gregory and Mead, Margaret. *Balinese Character: A Photographic Analysis*. New York: New York Academy of Sciences, 1942.

Battcock, Gregory. *New Artists Video*. New York: Dutton, 1978.

Baudrillard, Jean. *For a Critique of the Political Economy of the Sign*. St. Louis: Telos Press, 1981.

Bell, Daniel. *The Winging Passage*. Cambridge, MA: ABT Books, 1980.

Benjamin, Walter. *Illuminations*. New York: Schocken, 1976.

Berger, John. *Ways of Seeing*. New York: Penguin Books, 1977.

Berke, Joseph, ed. *Counter Culture*. London: Peter Owen Ltd. 1969.

Birdwhistell, Ray L. *Kinesics and Context*. Philadelphia: University of Pennsylvania Press, 1970.

Blanchard, Simon and Morley, David. Eds. *What is This Channel Fo(u)r?* London: Comedia, 1982

Bluem, William A. *Documentary in American Television*. New York: Hastings House, 1965.

Brecht, Bertolt. *Brecht on Theatre*. Trans. John Willet. New York: Hill and Wang, 1964.

Brown, Les. *Television: The Business Behind the Box*. New York: Harcourt Brace Jovanovich, 1971.

Cha, Theresa Hak Kyung, Ed. *Apparatus*. New York: Tanam Press, 1981.

Cole, Barry, Ed. *Television Today: A Close-up View*. Oxford University Press, 1981.
Chomsky, Noam. *Aspects of the Theory of Syntax*. Cambridge, MA: MIT Press, 1972.
Collier, John Jr. *Visual Anthropology: Photography as a Research Method*. New York: Holt, Reinhart & Winston, 1967.
Conrad, Peter. *Television: the Medium and it's Manners*. London: Routledge and Kegan Paul, 1982.
D'Agostino, Peter and Muntadas, Antonio. Eds. *The Un/Necessary Image*. New York: MIT/Tanam Press, 1982.
______________.and Thomas, Lew. Eds. *Photography: The Problematic Model*. San Francisco: NFS Press, 1981.
Davis, Douglas. *Art and the Future*. New York: Praeger, 1973.
____________.*Artculture: Essays on the Post-Modern*. New York: Harper and Row, 1977.
Davis, Douglas and Simmons, Allison. Eds. *The New Television: A Public Private Art*. Cambridge, MA: MIT Press, 1977.
Diamond, Edwin. *Sign Off: The Last Days of Television*. Cambridge, MA: MIT Press, 1982.
Eagleton, Terry. *Literary Theory*. Minneapolis: University of Minnesota Press, 1983.
Eco, Umberto. *A Theory of Semiotics*. Bloomington, IN: Indiana University Press, 1976.
__________.*The Role of the Reader: Explorations in the Semiotics of Texts*. Bloomington, IN: Indiana University Press, 1979.
Ellis, John. *Visible Fictions: Cinema, Television, Video*. Boston: Routledge and Kegan Paul, 1982.
Ellul, Jacques. *The Technological Society*. New York: Vintage Books, 1964.
___________.*The Technological System*. New York: Continuum, 1980.
Enzensberger, Hans Magnus. *The Consciousness Industry*. New York: The Seabury Press, 1974.
Epstein, Edward Jay. *News from Nowhere: Television and the News*. New York: Random House, 1973.
Fiske, John and Hartley, John. *Reading Television*. New York: Methuen and Co., 1978.
Foster, Hal. Ed. *The Anti-Aesthetic: Essays on Post-Modern Culture*. Port Townsend, WA: Bay Press, 1983.
Foucault, Michel. *Power/Knowledge: Selected Interviews and Other Writings*. New York: Pantheon, 1980.
Gans, Herbert J. *Popular Culture and High Culture*. New York: Basic Books, 1974.
Gardner, Howard. *Art, Mind, and Brain: A Cognitive Approach to Creativity*. New York: Basic Books, 1982.
Garnham, Nicholas. *Structures of Television*. BFI Television Monographs #1, London: British Film Institute, 1964.
Gerbner, George et al., Eds. *The Analysis of Communications Content*. New York: John Wiley, 1969.
Hanhardt, John G. Ed. *Nam June Paik*. New York: The Whitney Museum of American Art and W.W. Norton Co., 1982
Himmelstein, Hal. *Television Myth and the American Mind*. New York: Praeger, 1984.
______________.*On The Small Screen: New Approaches in Television and Video Criticism*. New York: Praeger, 1981.
Gitlin, Todd. *The Whole World is Watching: Mass Media in the Making and the Unmaking of the New Left*. Berkeley, University of California Press, 1980.
_________.*Inside Prime-Time*. New York, Pantheon, 1983.
Goffman, Erving. *Frame Analysis: An Essay on the Organization of Experience*. New York: Harper and Row, 1974
Gordon, George N. *Persuasion: The Theory and Practice of Manipulative Communications*. New York: Hastings House, 1971.

Innis, Harold. *The Bias of Communication*. Toronto: University of Toronto Press, 1951.
__________. *Empire and Communications*. New York: Oxford University Press, 1950.
Kaplan, E. Ann, Ed. *Regarding Television*. The American Film Monograph Series. Frederick, MD: University Publications of America, Inc., 1983.
Karnouw, Stanley. *Vietnam: A History*. New York: Viking Press, 1983.
Lacan, Jacques. *The Language of the Self*. New York: Delta, 1968.
Lambert, Stephen. *Channel Four: Television With a Difference?*. London: British Film Institute, 1982
Lessing, G. E. *Laocoön: An Essay on the Limits of Painting and Poetry*. Baltimore: John Hopkins University Press, 1984. (Originally published in 1766.)
Macdonald, Dwight. *Against the American Grain*. New York: Random House, 1962.
Mamber, Stephen. *Cinema Verite in America*. Cambridge, MA: MIT Press, 1974.
Marcuse, Herbert. *The Aesthetic Dimension*. Boston: The Beacon Press, 1977.
McLuhan, Marshall. *The Gutenberg Galaxy*. Toronto: University of Toronto Press, 1962.
__________. *Understanding Media: The Extensions of Man*. New York: McGraw-Hill, 1964.
Merrill, John C. and Lowenstein, Ralph L. *Media, Message and Men: New Perspectives in Communications*. New York/London: Longman, 1979.
Miller, George. *The Psychology of Communication*. Baltimore: Penguin, 1967.
Mills, C. Wright. *Power, Politics & People: The Collected Essays of C. Wright Mills*. London/New York: Oxford University Press, 1967.
Mosco, Vincent. *Pushbutton Fantasies*. Norwood, NJ: Ablex, 1982.
Newcomb, Horace. *TV: The Most Popular Art*. New York: Doubleday, 1974.
__________. Ed. *Television: The Critical View*. New York: Oxford University Press, 1976.
Nichols, Bill. *Ideology and the Image*. Bloomington, IN: Indiana University Press, 1981.
O'Connor, John E., Ed. *American History/American Television: Interpreting the Video Past*. New York: Frederick Ungar Publishing Co., 1983.
Ong, Walter J. *Orality and Literacy: The Technologizing of the Word*. New York: Metheun and Co., 1982.
Price, Jonathon. *The Best Thing on Television: Commercials*. Baltimore: Penguin, 1978.
__________. *Video-Visions: A Medium Discovers Itself*. New York: New American Library, 1977
Ricoeur, Paul. *Freud & Philosophy*. Trans. by Denis Savage. New York: Yale University Press, 1977.
Root, Jane. *Pictures of Women*. London: Pandora, 1984
Rosenberg, Bernard and White, David M., Eds. *Mass Culture: The Popular Arts in America*. New York: The Free Press, 1957.
Rosenthal, Alan. *The New Documentary in Action: A Casebook in Filmmaking*. Berkeley: UC Press, 1971.
Schiller, Daniel. *Telematics and Government*. Norwood, NJ: Ablex, 1982.
Schiller, Herbert I. *Mass Communications and the American Empire*. Boston: Beacon, 1970.
__________. *The Mind Managers*. Boston: Beacon, 1973.
Schneider, Ira and Korot, Beryl, Eds. *Video Art: An Anthology*. New York: Harcourt, Brace, Jovanovich, 1976.
Schramm, Wilbur, Ed. *Mass Communications*. Urbana, IL: University of Illinois Press, 1949.
__________ and Roberts, Donald F. *The Process and Effects of Mass Communications*. (Urbana, etc.), 1971.
Shamberg, Michael, and Raindance Corporation. *Guerilla Television*. New York: Holt, Rhinehart and Winston, 1971.

Sontag, Susan. *On Photography*. New York: Farrar, Straus and Giroux, 1973.
Shayon, Robert Lewis, Ed. *The Eighth Art*. New York: Holt, Rhinehart and Winston, 1971.
Sterling, Christopher, H. and Kittross, John M. *Stay Tuned: A Concise History of American Broadcasting*. Belmont, CA: Wadsworth, 1978.
Thomas, Lew. *Structuralism and Photography*. San Francisco: NFS Press, 1978.
Thomas, Sari, Ed. *Film/Culture: Explorations of Cinema in its Social Context*. Metuchen, NJ: The Scarecrow Press, 1982.
Venturi, Robert, et al. *Learning from Las Vegas*. Cambridge, MA: MIT Press, 1977.
Watkins, Peter. *The War Game*. New York: Avon Books, 1967.
Weiner, Norbert. *The Human Use of Human Beings: Cybernetics and Society*. New York: Avon, 1967.
Williams, Raymond. *Television: Technology and Cultural Form*. New York: Schocken Books, 1975.
—————————. *The Year 2000*. New York: Pantheon Books, 1983.
Williamson, Judith. *Decoding Advertising*. Boston: Marion Boyars Publishers, 1978.
Wittgenstein, Ludwig. *Philosophical Investigations*. Trans. by G. E. M. Anscombe, New York: MacMillian, 1968.
Wollen, Peter. *Signs and Meaning in the Cinema*. Bloomington, IN: Indiana University Press, 1972.
Worth, Sol. *Studying Visual Communication*. Philadelphia: University of Pennsylvania Press, 1981.
Worth, Sol and Adair, John. *Through Navaho Eyes: An Exploration in Study in Film Communication and Anthropology*. Bloomington, IN: Indiana University Press, 1972.
Wright, Charles R. *Mass Communication: A Sociological Perspective*. New York: Random House, 1975.
Youngblood, Gene. *Expanded Cinema*. New York: E. P. Dutton, 1970.

Periodicals

Afterimage. Rochester, NY: Visual Studies Workshop.
American Film: Journal of the Film and Television Arts. Washington, D.C.: The American Film Institute.
Art Com. San Francisco: Contemporary Arts Press.
Artforum. New York: Artforum International Magazine, Inc.
Cahiers du Cinéma. Paris: Edition de l'Etoile.
Film Library Quarterly. New York: Film Library Information Council.
Fuse. Toronto: Arton's Publishing Inc.
The Independent. New York: Foundation for Independent Video and Film.
Parachute, revue d'art contemporain. Montreal: Parachute, revue d'art contemporain.
Radical Software. New York: Gordon and Breach. (Out of print).
Screen. London: The Society for Education in Film and Television Ltd.
SEND. San Francisco: San Francisco International Video Festival.
Studies in Visual Communication. Philadelphia: Annenberg School of Communications
TV Magazine. New York: Artists' Television Network.
Video Texts. New York: Anthology Film Archives.
Videography. New York: United Business Pubications, Inc.

Contributors

JON P. BAGGALEY is a professor at Concordia University, Montreal, Quebec and co-author of *Dynamics of Television.*

ERIK BARNOUW is Professor Emeritus of Dramatic Arts, Columbia University and the editor of *The International Encyclopedia of Communications, to be published in 1988 by the Annenberg School of Communications and Oxford University Press.*

ROBERT BLY is a poet living in Minnesota. His books include The Light Around the Body, which won the National Book Award in 1968, *This Tree will be Here for a Thousand Years* and *News of the Universe.*

DEIRDRE BOYLE teaches at the New School for Social Research and Fordham University in New York. She is contributing editor to *Sightlines* and a Guggenheim Fellow. Her book, *Video Classics: A Guide to Video Art and Documentary Tapes* will be published by Oryx Press in 1985

JOHN CAREY teaches in the Interactive Telecommunications Program at New York University and is a consulting editor for *Studies in Visual Communications.*

PETER D'AGOSTINO is a professor of communications, Department of Radio-Television-Film, Temple University, Philadelphia. He is a television artist, a fellow at the Center for Advanced Visual Studies, MIT, Cambridge and a past fellow of the National Endowment for the Arts.

SUSAN DOWLING is director of the New Television Workshop, WGBH, and co-director of The Contemporary Art Television Fund in collaboration with the Institute of Contemporary Art, Boston.

STEVEN DUCK is a lecturer, Department of Psychology and co-editor of the SAGE Series in Personal Relationships at the University of Lancaster, England. He is co-author of *Dynamics of Television.*

JOHN FISKE is co-author of *Reading Television* and currently visiting professor at the University of Iowa.

HOWARD GARDNER is affiliated with Harvard Project Zero, the Boston University School of Medicine, and the Boston Veterans Administration Medical Center. He is a MacArthur Prize Fellow and author of *Art, Mind and Brain.*

MARTHA GEVER is editor of *The Independent,* the publication of the Foundation for Independent Film and Video. She is a past editor of *Afterimage* a publication of the Visual Studies Workshop, Rochester, N.Y.

PETER GIDAL is a London-based filmmaker and writer. His writings have been published in *Screen, Wide Angle* and *October* and many other journals. He is also the editor of *Structural Film Anthology,* BFI, 1975.)

TODD GITLIN is a professor of sociology and director of the mass communications program at the University of California, Berkeley. His books include, *The Whole World is Watching: Mass Media in the Making and Unmaking of the New Left* and *Inside Prime Time.*

JOHN HANHARDT is curator and head of the Film and Video Department at the Whitney Museum of American Art, New York. His books and catalogues include: *Nam June Paik* and *A History of the American Avant-Garde Cinema.*

JOHN HARTLEY is a lecturer in the Department of Behavioural and Communication Studies, The Polytechnic of Wales. He is co-author of *Reading Television.*

HAL HIMMELSTEIN is a professor in the Department of Television and Radio at Brooklyn College of the City University of New York. He is the author of *On the Small Screen* and *Television Myth and the American Mind.*

KATHLEEN HULSER is a journalist specializing in media and the arts. Her writing has appeared in *The New York Times, American Film* and *Afterimage.* She served as the editor of *The Independent* from 1982 to 1984.

JOANNE KELLY is a video artist and co-cirector of Video Free America, San Francisco. Her recent programs have been broadcast on KQED-TV.

BARBARA LONDON is video curator of the Department of Film at the Museum of Modern Art, New York. Her publications include articles on Japanese and Latin American video.

VINCENT MOSCO is a professor in the Department of Sociology, Queen's University, Kingston, Canada. He is author of *Pushbutton Fantasies* and is co-editor of *The Critical Communications Review.*

BILL MOYERS was Executive Editor of Bill Moyers' Journal, WNET, New York. His commentary now appears regularly on the *CBS Evening News.*

PAT QUARLES teaches in the Interactive Telecommunications Program at New York University.

ROBERT ROSEN is director of the ATAS-UCLA Television Archives.

DAVID A. ROSS is director of the Institute of Contemporary Art in Boston. In the past, he has served as video curator at the Everson Museum and deputy director of television and film at the Long Beach Museum of Art. He has written extensively about video art.

MARITA STURKEN is a free-lance video and film critic in New York who writes frequently for *Afterimage.* She is the author of the Circulating Video Library Catalog of the Museum of Modern Art.

JAMES WELSH is a professor of English, and co-founding editor of *Literature/Film Quarterly,* at Salisbury State College, Salisbury, Maryland. He is currently working on *Peter Watkins: A Guide to References and Resources* to be published by G. K. Hall and Company

SOL WORTH (1922-1977) was a professor at the Annenberg School of Communications and editor of *Studies in Visual Communication,* University of Pennsylvania. His book *Studying Visual Communication* was edited by Larry Gross.

Acknowledgements

"Hall of Mirrors" from *Dynamics of Television*, Gower Publishing Company, London, © John Baggaley and Steven Duck 1976. Reprinted by permission of the authors.

"The Functions of Television" from *Reading Television*, Metheun, London and New York, © John Fiske and John Hartley. Reprinted with permission of the publisher.

"Television Culture" from *On the Small Screen: New Approaches in Television and Video Criticism* by Hal Himmelstein, © Praeger Publishers, New York. Reprinted by permission of the publisher.

"Watching Television" © 1984 John Hanhardt. Printed with permission of the author.

"The Whole World is Watching" is a revised version of the introduction from *The Whole World is Watching: Mass Media in the Making and Unmaking of the New Left* by Todd Gitlin, © 1980 The Regents of the University of California. Reprinted by permission of The University of California Press.

"Margaret Mead and the Shift from 'Visual Anthropology' to the 'Anthropology of Visual Communication'" from *Studying Visual Communication* by Sol Worth, © Tobia Worth, 1981. Reprinted with permission of Tobia Worth.

"Cracking the Codes of Television: The Child as Anthroplogist" from *Art, Mind and Brain*, Basic Books, New York, © Howard Gardner. Reprinted by permission of the author.

"Interactive Television" © 1984 John Carey and Pat Quarles. Printed with permission of the authors.

"What is Videotex?" from *Pushbutton Fantasies* by Vincent Mosco, © 1982 Ablex Publishing Corporation, New Jersey. Reprinted by permission of the publisher.

"Ernie Kovacs: Video Artist" from *The National Video Festival Catalog, 1983* © Robert Rosen. Reprinted courtesy of the American Film Institute.

"Nam June Paik's Videotapes" by David Ross, from *Nam June Paik* edited by John Hanhardt © 1982 The Whitney Museum of American Art. Reprinted by permission of the publisher.

"Samuel Beckett's Ghost Trio" © Peter Gidal originally appeared in *Artforum*, May, 1979. Reprinted by permission of the author.

"Nuclear Consciousness on Television" © James Welsh, 1984. Printed with permission of the author.

"The Case of the A-Bomb Footage" © 1982 Erik Barnouw originally appeared, in a slightly different version, in *Studies in Visual Communication*, winter 1982.Reprinted by permission of the author.

"Guerilla Television" © 1984 Deidre Boyle. Printed with permission of the author.

"Meet the Press: On Paper Tiger Television" © 1983 Martha Gever originally appeared in *Afterimage*, November, 1983, Visual Studies Workshop, Rochester, NY. Reprinted by permission of the author.

"Poet at Large: A Conversation with Robert Bly" by Bill Moyers, © 1975 Educational Broadcasting Corp. Reprinted with permission of Bill Moyers and Educational Broadcasting Corp.

Photo Credits

cover: *The Bad Sister* by Mulvey/Wollen, photo credit: Keri Pickett; *Double You(and X,YZ)*, Peter D'Agostino, courtesy EAI; Ernie Kovacs from *EUGENE*.
page 2: *The ICA Kit, 1928*, courtesy the Bohle Company.
page 12: *Handing from The Austria Tapes* by Douglas Davis, courtesy EAI.
page 28: *Teletapes* by Peter D'Agostino, courtesy EAI.
page 42: *Media Burn*, Ant Farm, courtesy EAI.
page 56: *Reverse Television* by Bill Viola, photo credit: Kira Pirov, courtesy WGBH.
page 76: *An American Family*, Gary Gilbert (prod.), courtesy WNET
page 92: *Crossings and Meetings* by Ed Emshwiller, courtesy EAI.
page 104: *See it Now* (the first broadcast), Edward R. Murrow, courtesy CBS News,
page 118: *Media Ecology Ads* by Antonio Muntadas, courtesy of the artist.
page 142, 146-9: *Ernie Kovacs from Eugene*, courtesy Peter D'Agostino.
page 150: *TV Buddha* by Nam June Paik, courtesy MoMA.
page 164, 166-7: *Ghost Trio* by Samuel Beckett, courtesy Peter D'Agostino.
page 176: *The War Game* by Peter Watkins
page 188: Akira Iwasaki and Erik Barnouw, Tokyo 1972, courtesy of Erik Barnouw
page 202: *Four More Years*, TV/TV
page 214: *Paper Tiger TV: Herb Schiller Reading* The New York Times, photo credit: Vicki Gholsen, courtesy DeeDee Halleck.
page 262: *The Bad Sister*, Mulvey/Wollen
page 268: *The Lessons*, by Robert Ashley, courtesy the Kitchen
page 274: *Pop-Pop-Video: Kojak/Wang* by Dara Birbaum, courtesy EAI
page 280: *Synthesized Video Images* by Stephen Beck.
page 284: *Selected Works: 1974-6* William Wegman, photo credit: Kira Perov, courtesy MoMA.

* * *

Thanks to Teresa Bramlette and Ellen Cooper for their work on the production of the book; to Radha Mylapore, Larry Bansbach, and Don Cox for help in preparing the manuscript; to Laurie Zippay and Barbara London for assistance in obtaining photographs; to Carol Brandenburg; and to the many students who participated in my Television Aesthetics and Experimental Video courses at Temple University.

I also want to acknowledge my ongoing dialogues with Lew Thomas and Antonio Muntadas.

Tanam Press gratefully acknowledges the support of the National Endowment for the Arts and the Beards Fund.

Index(selected)

TANAM PRESS BOOKS

APPARATUS edited by Theresa Hak Kyung Cha. A collection of writings on film theory and practice, including works by Roland Barthes, Dziga Vertov, Jean-Louis Baudry, Maya Deren, Gregory Woods, Daniele Huillet, Jean Marie Straub, Thierry Kuntzel, Bertrand Augst, Marc Vernet and Christian Metz. 420 pages, sewn paperpack $12.95, hardcover $25.95.

FASSBINDER. A monograph on the late German filmmaker with essays by Peter Iden, Ruth McCormick, Yaak Karsunke, Wolfram Schutte, Wilfried Wiegand and Wilhelm Roth. Also included is a surprising interview with Fassbinder and an extensive annotated filmography. 93 photographs, 256 pages, sewn paperback $8.95, hardcover $16.95

SCREENPLAYS by Werner Herzog. The original narrative treatments for the films, *Aguirre, the Wrath of God* and *Every Man for Himself and God Against All* and a record of the dialogue from *Land of Silence and Darkness.* 204 pages, hardcover $12.95.

OF WALKING IN ICE by Werner Herzog. The journal of Herzog's walk from Munich to Paris in the winter of 1974 to find Lotte Eisner. 96 pages, hardcover $10.95

SATYAGRAHA: M.K. GANDHI IN SOUTH AFRICA by Constance DeJong and Philip Glass. This volume includes the text, historical background and photographic documentation of one of the key music-theater events of the 20th century. 80 pages, sewn paperback $5.95, hardcover $12.95

WILD HISTORY edited by Richard Prince. A collection of contemporary writing from New York City. Contributors include: Spalding Gray, Constance DeJong, Kathy Acker, Tina Lhotsky, Reese Williams, Anne Turyn, Peter Nadin, Roberta Allen, Glenn O'Brien, Gary Indiana. Richard Prince, Sylvia Reed, Robin Winters, Collins/Milazzo, Cookie Mueller, Lynne Tillman, Paul McMahon and Wharton Tiers. 256 pages, paperback $10.95

DICTEE by Theresa Hak Kyung Cha. A series of nine narratives with each of the Nine Muses identifying each of the sections. The narratives trace names, events, and histories of existing persons, individual personages in history and other fictitious characters embodied in nine female voices. Each is an evocation of the past through speech, through the research of language that may open avenues to Memory, to the elemental process of recollection. 176 pages, sewn paperback $6.95, hardcover $13.95

PALABRA SUR edited by Cecilia Vicuña. A collection of bilingual editions presenting the work of poets from Latin America. Please write for further information.

ORDERING INFORMATION: All titles are available by mail order; prices include postage and handling; send check or money order to Tanam Press, 40 White Street, New York, NY 10013.